To Wang Qin

Foreword

Achieving improved health is an overall goal of the health services system. Only on the basis of appropriate and equal health care financing can the goal be achieved. In urban areas of developing countries health insurance can be readily established on an employment basis. In contrast, in rural areas the health insurance is more problematic and institutional innovations are needed to address the problem of high transactions costs. Community-based health insurance scheme (CBHIS) has been recognized as a promising institutional arrangement and played an increasingly important role in the health system of many low and middle-income countries in the past decade.

China's rural cooperative medical scheme (RCMS) beginning in 1955 was one of the first large nationwide community-based health insurance scheme to fund and organize prevention, primary care and secondary health services for its rural population. It caused China to achieve a great deal of health improvement from the 1950s to the early 1980s at a low-income level. However, it broke down during the transition from the centrally planned to market-oriented economy. Only 7% of rural Chinese were members of RCMS in 1998. To date, many efforts of China's government and communities to expand private health insurance and RCMS have proved ineffectual.

This current dissertation research by Yuansheng Jiang is part of a Germany-China collaborative research on health insurance and development in developing countries. The research in rural China is significant for China, providing China's government, private insurers and communities not only with some effective ways to expand demand for private health insurance (PHI) and RCMS but also with knowledge on strengthening the rural households' ability to manage health risk with informal insurance arrangements.

Yuansheng Jiang has provided especially innovative and noteworthy research by explicitly including the concept of the extended family in his public and private insurance studies. He quantified the insurance functions of the extended family in terms of pool size, composition and transfers. The existence of extended family and its frequent reciprocal transfer for health care financing in rural China facilitated coping with the decline in RCMS. However, the demographic changes due to migration, low birth rates and aging put this insurance system at risk. Institutional innovations in health care financing with public and private systems need to be urgently explored and scaled up in coming years.

Prof. Dr. Franz Heidhues
Center for Tropic Agriculture
University of Hohenheim

Prof. Dr. Joachim von Braun
International Food Policy Research Institute (IFPRI)
Washington, DC

Abstract

It is widely recognized that there exists a vicious cycle between poverty and poor health, and government interventions in the health sector are indispensable to break the cycle. Most governments have been attacking poverty by enhancing the human capital of the poor through provision of health services. However, developing countries have been constrained by the limited fiscal revenue. Thus, financial resources must be mobilized by a set of appropriate instruments. Health insurance is a highly practical instrument. However, it is problematic and an institutional innovation is needed in rural areas. Community based health insurance scheme (CBHIS) has been recently recognized as a promising instrument to apply in rural areas. Rural cooperative medical scheme (RCMS) is one form of CBHIS in China. This study tries to build up the knowledge on strengthening the rural population's ability to protect themselves against costs of illness by exploring ways to expand demand for private health insurance (PHI) and RCMS without impairing the role of the extended family.

A case study in rural China shows that both the PHI and RCMS are promising health care financing mechanisms. With respect to PHI, the rural households benefit from it by having costs of catastrophic diseases of their only-kids insured. The estimation of a binary probit model suggests that income, school attendance, and competition among suppliers are the most responsive factors to boost the demand. RCMS is a pro-poor insurance arrangement as it encourages the ill to report illness and seek care actively. Both the analyses on care-seeking behavior and on willingness to pay suggest that to raise the enrolment rate, RCMS should be improved by introducing a high coverage of both outpatient and inpatient services and a strict fund management. Although the rural community has an active role to play in establishing RCMS, the government subsidies, knowledge and personnel support are indispensable.

The informal insurance arrangements should not be neglected. The extended family, as a typical informal insurance arrangement in China, plays a major role in coping with risks in rural areas. A survey on its basic features, operational modes and the estimation of its possible impact on formal health insurance were conducted. Results show that the average size of extended family is nine households; the most important financial donors are brothers and sisters. The respondents' households received 16 times (a worth of 1627.1 Yuan) due to illness in the past. Of illness episodes in last year, 22 percent had to seek financial aids within the extended family. The regression analysis does not offer any evidence in favor of the hypothesis that the function of the extended family adversely affects the willingness to pay for the hypothetical RCMS.

Table of Contents

List of Tables

List of Figures

Abbreviations and Acronyms

ADB	Asian Development Bank
AMR	Adult Mortality Rate
BL	Binary Logit
CASS	Chinese Academy of Social Science
CBHIS	Community-Based Health Insurance Scheme
CLIC	China Life Insurance Company
CPC	Communist Party of China
CV	Contingent Valuation
GDP	Gross Domestic Product
GHIS	Government-employee Health Insurance Scheme
GNP	Gross National Product
HDI	Human Development Index
IMR	Infant Mortality Rate
LHIS	Labor Health Insurance Scheme
LE	Life Expectancy
MMR	Maternal Mortality Rate
MNL	Multinomial Logit
MOH	Ministry of Health
NSB	National Statistics Bureau of China
NSRHS	National Survey on Rural Health System
OLS	Ordinary Least Square
PF	Private Facilities
PHI	Private Health Insurance
RCMS	Rural Cooperative Medical Scheme
HISRS	Health Insurance Survey in Rural Sichuan
SSB	Sichuan Statistics Bureau
TFR	Total Fertility Rate
THC	Town Health Center
U5CMR	Under 5 Child Mortality Rate
UNDP	United Nations Development Program
VHS	Village Health Station
WTJ	Willingness to Join
WTP	Willingness to Pay
ZEF	Center for Development Research

Acknowledgments

This doctoral thesis would not have been possible without the strong support of a number of people and organizations. First of all, I would like to express my sincere thanks to Prof. Dr. Joachim von Braun, director general of the International Food Policy Research Institute (IFPRI), for his invaluable guidance and encouragement throughout my research, and I would like to thank him for his many inspiring and insightful comments on my dissertation, and especially I am greatly indebted to him for having offered me the opportunity to participate in the International Doctoral Study Program for Development Research at the Center for Development Research (ZEF), University of Bonn. I would also like to acknowledge Prof. Dr. Michael-Burkhard Piorkowsky for taking the role of second supervisor.

I am very much grateful to the China Scholarship Council (CSC), on behalf of the Ministry of Education, for their financial support for the first-year doctoral study. I am also very much grateful to the German Federal Ministry of Economic Cooperation and Development (BMZ) for funding the field research. Special thanks go to ZEF for its further financial support. I am highly grateful to Sichuan Agricultural University and its Faculty of Economy and Management for their generous material assistance.

Colleagues and good friends from ZEF, Institute for Economics of the Chinese Academy of Social Science and Sichuan Agricultural University have provided me with a great deal of advice and support. For the comments on my work, I would like to thank Dr. Abay Asfaw, Johannes Juetting, and Prof. Dr. Zhu Ling; Special thanks to Dr. Dong Hengjin, Department of Tropical Hygiene and Public Health of University of Heidelberg, for acting as discussant for my proposal and one progress report; for the useful discussion, I am indebted to Isaac Osei-Akoto, Dr. Peter Wehrheim, Dr. Zhou Jizhong and Dr. Zhu Qiuxia; for general advice, support and proof-reading, many thanks go to Daphne Fritze, Dr. Li Zhaohua, Hana Ohly, and Benjamin Kofi Nyarko. Many acknowledgements go to Dr. Günther Manske for his careful coordination throughout the study and field research. I would also like to send my hearty indebtedness to Prof. Wen Xintian, associate Prof. He Linchun and Xiao Hongan for providing me with convenience in study and survey.

My acknowledgement further extends to enumerators for their careful work, interviewees, health officials and managers in insurance companies for their willingness to talk with us. My good friend, Wang Liming, is worth of my particular indebtedness. Without his coordination, the organization survey would not have been a fruitful one.

Finally, my sincere thanks go to my dear wife, Wang Qin and my son for their love, understanding, and moral support throughout the years. I also express my hearty thanks to my parents, parents-in-law, grandmother, aunt, uncles, brothers and sisters for their understanding and spiritual support.

1. Introduction

1.1 Background and problem statement

Despite huge efforts made to improve human health, a vast number of people on our planet are not only subjected to a series of deadly diseases: AIDS, malaria, tuberculosis, and viral hepatitis, etc., but also are suffering and dying due to lack of safe water, adequate sanitation and basic medicines. Every year in the developing world, 12.2 million children under age 5 die, most of them from preventable causes, preventable, in many cases, for just a few US cents (WHO 1995). In this sense, poverty is the world's most fatal disease.

Even though the number of absolute poor people (that is, living on less than $1 per day) declined between 1980 and 1998 by at least 200 million, the population stricken by poverty remained as high as 1.2 billion until 1998 (World Bank 2003). The poor health not only imposes financial burden on households but also undermines the productivity of the laborers, particularly for rural people who earn their basic living through physical labor (Dickson 1959). Thus, poverty is the consequence of poor health (Farmer 1999; Hamoudi and Sachs 1999). This is a vicious cycle of poverty, which is difficult to escape for those who are being trapped in it. There is an imperative need to break the vicious cycle with an appropriate health intervention.

Apart from the consideration of economic consequences, from the perspective of ethic it is also necessary to provide sufficient and universal primary health care for all. The Declaration of Human Rights and the Constitutions of most countries ensure the rights of all human beings to primary health care. A great many of governments and international organizations have been attacking poverty by enhancing the human capital of the poor people through provision of efficient and accessible health services. However, health reforms in developing countries have been universally constrained by the limited governmental revenue. In order to improve the health status of disadvantaged people, additional financial resources must be mobilized into the health sector by a set of appropriate policies or instruments. Health insurance is a highly practical instrument, which governments can utilize to relieve them from the high cost of providing subsidies for health care, thereby releasing funds for public health schemes and preventive and primary services that benefit the poor (Petretto 1999; Bloom 1998; Waters 2000; Liu *et al.* 1999; Shaw and Griffin 1995). This is also a useful instrument to build capacity of the poor to cope with health risk, given that the poor are less well insured both in income and in consumption (Jalan *et al.* 1999) and are liable to health shock.

Health insurance has three main functions: i) pooling and spreading the risk of illness; ii) encouraging the use of health care in the case of illness; iii) improving the equity in access to health care and preventing the poor and vulnerable from being trapped in poverty. Health insurance in urban areas is readily established on the basis of employment. In rural areas, however, the population in most developing countries is institutionally difficult to be covered by governmental health insurance scheme (Shaw and Griffin 1995) and an institutional innovation is needed. Community based health insurance scheme (CBHIS) has been recently recognized as a promising instrument to apply in rural areas of developing countries. CBHIS is in fact not a new institutional arrangement. As early as in the 1950s, China initiated a health risk sharing arrangement in its rural areas called rural cooperative medical system (RCMS), which had similar, if not identical, institutional, organizational and managerial attributes, compared with today's CBHIS. Then RCMS was rapidly constructed in the overwhelming majority of villages. In the late 1970s, the enrolment rate of RCMS reached a historic peak of over 90 percent. Although it was not without problem, it significantly contributed to China's many accomplishments in health sector over the past decades, especially before the economic reforms. The accomplishments encompass a fast improvement in health status, a broadening physical access to primary health care, and a support of important public health measures (World Bank 1997).

With the deep social and economic change occurring since the end of the 1970s, RCMS broke down nationwide in spite of the rapid and persistent economic growth. China's health sector has confronted many problems. According to WHO (1996), health inequality, in terms of health status, health care utilization and health care financing, between the urban and rural population, is the major issue of much concern in recent years. The basic and logical relationship among these three aspects is that unequal financial access to health care brings about unequal health care utilization, and unequal health care utilization leads to unequal health status.

It is alarming to note that a gap in health status between the urban and rural populations in transitional period is widening (Liu *et al* 1999). In the rural areas, signs of health improvement are few (Bogg *et al* 1997). A survey on health services and health status in 300 poverty-stricken counties (12% of all counties in China) appears to confirm this relation (Liu *et al* 1996). Despite the low economic growth rate, the poverty-stricken counties have experienced steady economic growth after economic reform. Per capita GDP in real term increased in the late 1980s. Nevertheless, the health status of people in poverty-stricken counties shows signs of deterioration. Median infant mortality rate (IMR) in the surveyed poor counties increased from about 50 to 72 per 1000 live births during the 1980s.

Similar trend is observed in health care utilization. There was a relative decline in utilization of hospital services by the rural population from the mid-1980s to the early 1990s. Between 1985 and 1993 annual inpatient days per 1000 urban residents increased by 12.8 percent, whereas the rural population experienced a decline of 10.3 percent in inpatient services. A further study in the period from 1993 to 1998 showed that there was a slight decrease in the inpatient services for the urban and a close to constant level in inpatient admissions for the rural. Concretely speaking, the rural-urban ratio increased from 0.59 in 1993 to 0.61 in 1998 (Gao and Qian *et al* 2002). In spite of the narrowing gap between the rural and urban population, the stagnancy of the rural inpatient admission rate is unexpected.

Various factors are responsible for this dismal performance in rural areas, especially in the poor areas. Many researchers agree that apart from the policy of public health spending, which is heavily biased toward the urban, a reduction in medical resources and the collapse of three-tier health system caused by the breakdown of RCMS are the most important reasons. For example, the World Bank (1997) pointed out that the under-five child mortality rate (U5CMR) in China declined steadily until the early 1980s and then stagnated until 1991. The decline and stagnation of U5CMR coincided with the breakdown of RCMS in time. Experience from other countries suggests that U5CMR need not plateau as China's did in the late 1980s (as measured by census and survey data). Liu *et al* (1999) argued that the absence of health improvement is associated with changes in health care financing and organization, including dramatic reduction of insurance coverage for the rural population and relaxed public health.

In recognition of the failure of market for the health insurance, Chinese governments endeavored to intervene in the health insurance market. In 1993, the central committee of the Communist Party of China (CPC) proposed to develop and optimize the rural cooperative medical system in an important resolution, titled *Resolution on a few Issues of Construction of the Socialist Market Economy*, followed by a large-scale pilot research in 14 counties of seven provinces. In 1997, based on *Resolutions on Health Reform and Development* proposed by the central committee of CPC and the State Council, the Ministry of Health set an ambitious agenda that RCMS would be restored to the historical peak of 90 percent by 2000 (Chen *et al* 2001). The provincial governments set corresponding agendas and adopted some measurers. However, efforts have proven ineffectual. The enrolment rate of RCMS remained below 10 percent in 1998 (MOH 1999), and some 11 percent in 2000 according to a National Survey on the Rural Health System (NSRHS) conducted by Chinese Academy of Social Science (CASS).

The critical issue appears to be that the government lacks essential information on demand side. For instance, the determinants for a community to have CBHIS have not been identified; the determinants of participating in private health insurance are not known to both governments and insurance companies; the role of private health insurance in mobilizing resources is unclear. The governments usually do not know what level of premium and coverage of RCMS is preferable for the rural prospects in different regions, and what role the local communities can assume. Thus, the governments are unable to propose and execute appropriate and effective policies in order to expand demand of households for health insurance in the short run.

Therefore, there is an urgent need to identify the determinants for buying health insurance policies or joining RCMS by estimating the demand function for health insurance. Among many factors that affect demand for health insurance, the substitutive effects of informal insurance mechanisms must be highlighted. In the long history of without formal health insurance arrangements, some informal insurance mechanisms have been formed spontaneously, such as drawing on savings, selling assets, receiving transfers from their relatives and social support networks, borrowing money in local credit markets, and even the dropout the of school children (Zeller 1999; Gertler and Gruber 2000) and have played a major role in coping with various risks. It has been argued that the introduction of formal health insurance might crowd out the informal insurance mechanism, and offset the welfare gain to a certain degree (Coate 1993; Atim 1999; Morduch and Sicular 1999). In contrast, the informal insurance mechanism may have a negative influence on the formal health insurance market. Private transfers within the extended family (including good friends) are common in developing countries, such as China, as noted by some economists (Morduch 1999; Fafchamps and Lund 2002). Member households conducted reciprocal transfers frequently, and the reciprocal transfers mainly took the forms of labor, gifts, cash and the like. There is quite few empirical data on the scale of risk pool, the magnitude and frequency of the private transfers. Furthermore, the impact of this informal insurance mechanism on formal insurance has not been investigated in the literature.

Alleviation of poverty in China can only be achieved by first reducing the poverty in rural areas. The economic system in China is in transition, from a centrally planned to a market-oriented one. The rapid change of the institution not only caused the collapse of rural social security systems, but also greatly reduced the provision of public goods. The rural population has never been so vulnerable to various drastic changes. With 65 percent of the population living in the rural areas and 75 percent of the poor belonging to the rural population, Chinese rural areas deserve special attention. A majority of studies on health insurance demand concentrated mostly on urban areas (e.g., Henderson *et al* 1994; Costa *et al* 1999). Yet, little research has been conducted to investigate health insurance demand and health risk management behaviors in rural areas.

1.2 Research objectives

The objective of this thesis is to fill the wide research gap in empirical study of demand for health insurance and health risk management in rural China. Based on the findings and policy recommendations governments and local communities would be able to stimulate demand for health insurance by improving some responsive factors and suppliers would be able to provide health insurance products that rural households demand largely and urgently. The specific objectives are:

- To investigate both the observed and the potential demands for health insurance quantitatively and identify the determinants of demand;

- To explore the impact of health insurance on health care seeking behavior in rural China;
- To assess the performance of individual rural cooperative medical scheme and the role of RCMS in health improvement;
- To observe and analyze risk coping behaviors of rural people and investigate the impact of private transfer within extended family on demand for formal health insurance schemes.
- To compare the extended family with the private insurers, the rural cooperative medical schemes from the perspective of institutional, organizational and managerial characteristics.

1.3 Fundamental assumption and working hypotheses

Although market failure may result from the inadequacy of both supply and demand, it is reasonable to believe that the inadequate demand is dominant in the world today with excessive productive power. More specifically, the health insurance products are produced and sold simultaneously. Once a product is actuarially designed, the marginal cost of production is negligible. The fundamental assumption of the study is that the low coverage of health insurance in rural areas is mainly due to inadequate demand for health insurance and for community based health insurance scheme.

The main working hypotheses of this study comprises as follows:

- Formal health insurance is more capable of coping with financial costs of illness but need not be more efficient and effective than other informal insurance arrangements;
- The richer group prefers to use the health insurance scheme to cope with health risk;
- Existing informal health risk coping mechanisms (e.g., the extended family) has a negative impact on purchase of private health insurance policies and on participation in community based health insurance schemes.
- The extension of formal health insurance in rural areas depends not only on the effort of the governments but also on the deep involvement of local communities and active participation of private insurers.

1.4 Study area and data sources

To achieve the five stated specific objectives, a comprehensive research was conducted in the People's Republic of China. The reasons underlying its selection are as follows: In the first place, China was the first large nation in the world to develop community health care financing schemes nationwide, which is called rural cooperative medical scheme (RCMS) in China and organizes delivery of preventive, primary care and secondary health services as well as their financing. However, the market-oriented economic reform launched at the end of 1970s caused a rapid collapse of RCMS. China's relatively successful experience in extending health care coverage to its rural population in the context of fairly low productivity and the subsequent rapid breakdown invoked interests throughout the developing world in the potential role of community health care financing (Abel-Smith and Dua 1987). Second, of the largest population and largest rural portion, China has acquired quite satisfied health outcomes corresponding to its level of gross domestic production (GDP). Finally, besides the rural cooperative medical system, the existence of private health insurance and some informal insurance arrangements makes it possible for a comparative research among the three options. Nevertheless, China is a big country with substantial social and economic difference among different regions and provinces; at the same time, one always faces the problem of time and resource scarcity. Therefore, putting the study in the context of the whole economy, the study focuses on a case study in Sichuan province, western China, which not only lags behind the eastern regions in terms of the social and economic development but is also home to a number of poverty-stricken people.

The study is based on a Health Insurance Survey in Rural Sichuan (HISRS) by the author in 2001 and a detailed National Survey on Rural Health System (NSRHS) in China conducted by Chinese Academy of Social Science (CASS) at the end of 2000. The national data set is used to estimate determinants for a community to have health insurance scheme and the impact of CBHIS on health outcome. The HISRS collected the data of households and insurance suppliers and it is the main data source of the econometric analysis of health insurance demand.

1.5 The structure of the thesis

The remainder of the thesis is organized as follows. Chapter 2 addresses major issues on the relationship between the health risk, health insurance and rural poverty by summarizing the existing literatures. In this chapter, firstly, health risk is defined and the scale of health risk is measured by selecting appropriate health indicators. Secondly, causal-effect relation between health risk and rural poverty are explored from many facets. Thirdly, the potential benefits of health risk coping or insurance are enumerated from perspectives of households and society respectively. Finally, the rationale of provision of health insurance by the governments and communities in rural areas is revisited.

Chapter 3 offers an overview of health care system in China's socio-economic setting. Firstly, this chapter examines the profile of China briefly, describing the physical, demographic, cultural, social, and economic aspects. Secondly, China's health care system is expatiated from facets of health care delivery, health care financing system as well as the reform process in the past two decades. The supply of health insurance (private health insurance and rural cooperative medical scheme) is discussed in the section of health care financing system. Finally, the profile of the study province, Sichuan, is described briefly.

Chapter 4 attempts to estimate demand for health insurance and its impact on health care seeking behavior. This chapter argues that with the rapid development of private health insurance in last decade, private insurers play an increasingly big role in providing health security for the rural population. First of all, determinants of demand for private health insurance are identified and the elasticity of income is computed based on the household survey data collected in Sichuan province. Then, determinants of establishing rural cooperative medical scheme (RCMS) in towns are explored. Finally, the impact of health insurance in general and private health insurance in particular on health care seeking behavior is described and econometrically analyzed.

Effort is made in Chapter 5 to assess the performance and the role of RCMS and to estimate the potential demand of rural households for the hypothetical RCMS through *contingent valuation* (CV) method. The performance of RCMS in China is assessed through 4 aspects: examining the existing literatures, analyzing health outcomes before and after the economic reforms, between counties with and without RCMS and presenting a case study in Sichuan province. In order to re-initiate RCMS, the baseline information on willingness to join and to pay for a hypothetical and improved rural cooperative medical scheme is required for sound and wise decision-making. The descriptive and econometric analysis on the basis of HISRS dataset is made to serve the purpose.

In chapter 6 coping strategies with economic costs of illness and the role of the extended family in risk coping is discussed. Although direct costs of illness inclusive of monetary and time costs of illness is moderate, compared with other developing countries such as Ethiopia and Burkina Faso, the proportion of direct costs of illness in household income is sufficiently high to affect the normal food consumption and may induce poverty in some cases. However, whether costs of illness cause poverty depends to a large extent upon the capacity of coping with the costs. This chapter also highlights the role of the extended family in coping with costs of illness. Specifically, the basic attributes of the extended family, e.g., the size, the frequency and amount of transfer, are measured in the first place; secondly, the effect of private transfers on income redistribution is explored by comparing income of respondents before the transfer with those after the transfer. Finally, the impact of the functioning extended family on demand for health insurance and willingness to pay for the hypothetical RCMS is analyzed by re-estimating the demand function considering the extended family.

The thesis concludes with chapter 7, in which main findings are summarized and policy recommendations are proposed on the basis of the previous analyses.

2. Major Issues in Health Risk and Rural Development

Although risk may be positively related to gain in business, people still like security rather than risk[1], because the adverse consequence of risk may go beyond the capacity of individuals, households or enterprises to withstand.Risk is not only the matter of individuals, households or enterprises, but it also hampers an economy's ability to accumulate, innovate, and develop. Risk, thus, contributes to creating a vicious circle of poverty. Health risk is not an exception. A proper understanding of the interaction between health risk and poverty is essential to identify the obstacle to growth and development in poor rural societies.

2.1 Health risk and rural poverty and development

2.1.1 Definitions and classifications of health risk

The terms of "risk" and "uncertainty" are usually intertwined with each other in literatures and there is a need to make a distinction. In line with common usage, uncertainty refers to imperfect knowledge and risk refers to uncertain consequence, particularly exposure to unfavorable consequence (Hardaker *et al* 1998). In addition, there are two types of risk, namely, speculative risk and pure risk. Speculative risk occurs when a person invests money in a business or in securities such as stocks. The investor gains profit in return for bearing the risk of loss. In pure risk there is no possibility of gain. There is only the possibility of no loss, or a loss due to unforeseen and unexpected hazards beyond the control of the one for whom the risk exists. This is the kind of risk that insurance is designed to cover.

Throughout this thesis risk is defined as "outcome of uncertainty". This definition includes two generally used interpretations, namely, risk as the chance of bad outcome and risk as the variability of outcomes (Young 1984). Accordingly, health risk is the chance of illness and the undesirable outcome resulted from the illness, such as the financial loss due to care seeking, time loss and the body suffering as well. Health risk can be classified into various types according to different criteria.

High and low frequency health risks: Risks can be classified into high- and low-frequency risks based on the frequency of risk events. Minor diseases occur frequently, for example, morbidity rates of acute pharyngitis coryza and influenza in the past two weeks are 42 cases per 1000 persons and 15 cases per

[1] According to psychologist Maslow's theory on "human hierarchy needs" that man has five levels of needs and the first and the second level of need are physiological needs and safety needs respectively.

1000 persons for rural Chinese, respectively (MOH 2002). Severe diseases occur rarely, for example, cerebrovascular disease occurs at a rate of 1.7 per 1000 for rural Chinese (MOH 2002). However, this disease is more dangerous. Interestingly, the high-frequency health risk usually is accompanied by low intensity of financial loss and suffering while the low-frequency health risk has high intensity of loss. Thus, it is easier to deal with small but frequent health problems than with large but infrequent health shocks.

Covariate and idiosyncratic health risks: Health risk can also be classified into covariate and idiosyncratic risks based on the spreading scope of health risk across individuals. The covariate health risk refers to the risk factor that affects a group of people, for example, influenza, virus hepatitis and sexually transmitted diseases. The idiosyncratic health risk affects only isolated individuals, cancer and heart troubles, for example, are in this class of risk.

Economists also divide risks into *utility* and *income* risks based on the extent to which individuals are affected. Utility risk refers to the risk factors that directly affect individuals' welfare or utility; income risk refers to the risk factors that affect income and wealth only (Faffchamps 1999). Yet, the boundary between the two is practically blurry in the case of health risk. Disease or injury affects the welfare directly and is a form of utility risk. In addition to the welfare loss, the ill or injured will incur income or wealth loss as long as they seek care. In this sense, the disease or injury can be considered income risk. The classification is especially instructional when measuring the costs of illness because utility risk such as pain and suffering used to be ignored and only income risk was addressed.

2.1.2 Measuring health risk in an economy

The uncertain health status imposes a number of risks on individuals, the most obvious is the loss of health itself, as well as the risk of incurring large financial costs associated with medical treatments aimed at maintaining or improving health. In addition, the health deterioration can reduce an individual's productivity so that the person faces the risk of lost wages or less yields from fields. Finally, individuals may be less able to enjoy other forms of consumption (e.g., participation in sports) because of poor health, or they may suffer the psychological and emotional trauma associated with physical deterioration.

Although individuals are uncertain about the exact size and occurrence of health risk they face, for a large sum of people it is predictable on the basis of law of large number, e.g., all the population or sub-population in an economy. Therefore, can the existing health indicators measure health risk? Health indicators are used for assessing a population's health status, monitoring implementation and outputs of a program, and evaluating the effectiveness and impact of a program (WHO 1998). Thus, the health indicators for a particular group of people are actually average health risks these people are facing. Health indicators can be divided into three categories: *morbidity, mortality* and *longevity,* each has its respective implication for health risks. The commonly used indicators for morbidity comprise self-reported two-week morbidity, bed-day due to illness and epidemics morbidity, which convey the information on both income risk and utility risk. Mortality indicators are widely composed of infant mortality rate (IMR), maternal mortality rate (MMR), population mortality rate (PMR), pregnant maternal mortality rate, and under-5 child mortality rate (U5CMR). Hence each indicator is used as a proxy for mortality of a specific subpopulation. From individual's perspective, this represents the most serious utility risk; from state's or family's perspective, this also means income risk, especially for those early deaths, because much has been invested in those who died without fulfilling their social obligation. Compared with the mortality, morbidity is less accurate because apart from the difficult definition of some diseases, a large number of patients have never entered health facilities. As a consequence, the statistic system on morbidity can hardly reflect the real situation. In contrast, the mortality is more credible because death in general is not being neglected. Moreover, most countries have established a complete official mortality registration system. Life expectancy is the scientific index for the longevity assessment calculated on the basis of population mortality. This illustrates health risk from the opposite side. Life expectancy is a very useful indicator, through which one can predict a person's potential longest age and arrange his or her lifetime personal activities. Information as such can be used by the state (or market) to allocate proportionate facilities and corresponding security for the aged.

2.1.3 Links of health risk to rural poverty and development

Morbidity, malnutrition and vulnerability to shocks

Health risk, using morbidity rate as an indicator, is related to both utility decrease and income loss (Fafchamps 1999). There are generally two possible outcomes for everyone who is sick: one is recovery from the disease and the other is death. Of course, death is dreadful for the dead themselves but it also results in much loss for the family and society, especially when one dies early[2]. For the family, a premature death may bring sorrow and suffering. It is even more frustrated and sorrowful that the losses are occasionally irreversible because the parents are too old to give birth to baby to compensate. The early death of adult breadwinner may cause a family into poverty. For a society, they are skilled technicians and professionals and their deaths may bring about gigantic economic loss to their working units. Obviously, early death has immeasurable negative externalities. So both the society and family have strong incentives to prevent early death. However, many governments and families in developing countries fail to translate the strong incentives into effective actions because of uptight revenues.

---

[2] Early death refers not only to the death of infants, children and adolescents, but also to death of adults. The former can also be called premature death

There may be intermediate consequences of morbidity rather than dying or recovering. Stunting, a serious deficit in height, is one of them. Stunting is generally believed to be the result of some morbidities and improper nutrition during childhood. Stunting can reduce an individual's body size and strength. Short size has empirically been associated with lower productivity in farming and livestock activities (Foster *et al* 1996; Fafchamps *et al* 1998). Furthermore, disease can similarly reduce someone's ability to cope with subsequent health shocks. Consequently, inability to cope with health risk at a moment in time leads to increase in vulnerability in the future.

By increasing the vulnerability of individuals or groups to health and income shocks, poverty in turn leads to high morbidity and a higher incidence of stunting, chronic poor health, and famines. These undoubtedly have an adverse effect on the stock of human capital and talent in a given society. They also disrupt the families and make it difficult for children to receive essential nutrition and adequate education. A proposition can be made that poverty itself leads rarely to death or disease except in cases of extreme poverty; the proximate cause of poor health or death is always an identifiable income or health shock. It is not poverty that kills or incapacitates the rural poor directly, but the vulnerability to shocks that is generated by poverty (Faffchamps 1999). Thus, the provision of sufficient nutrition and timely treatment for the sick helps lower subsequent vulnerability to shock and might be a quite efficient and effective instrument to combat poverty in the short term.

Morbidity rate and school attendance

Combating poverty through enhanced human capital has been receiving increasing attentions globally (World Bank 1980). Education, health and nutrition are three major components of the formation the human capital. The contribution of education is especially significant. Microeconomic empirical works suggest that, in poor areas, the return to education is highest in non-farm activities (e.g., Fafchamps *et al* 1998; Lavinas 2002; Angrist *et al* 1991). There is evidence which indicates that education might also contribute significantly to farm activities. For example, when faced with new agricultural technologies, the better-educated farmers are usually open and learn with little difficulties (Philips 1987). Poor people's capacity to invest in education is, thus, an important factor underlying the long-term success in developing rural areas.

Yet, how does health risk affect the school attendance? Health risk imposes an influence on household head's decision to invest in education in two paths: firstly, high mortality rate induces the household head to have more children than are anticipated to guarantee a minimal number of offspring. The increased number of children will compete against each other for the limited resources in the family. As a result of competition, there are always some kids who cannot be sent to school. It is justly believed that girls are usually disadvantaged in comparison with boys. Secondly, kids have to drop out of school temporarily due to poor health of themselves or of their sponsors. It is also quite likely that the temporary dropout can become a permanent one because their academic performances become worse and they lose the confidence to catch up with their classmates or have to repeat a grade, which will certainly leads to extra costs of education. Children's time is valuable for the rural poor because children often contribute to household work, participate in livestock and farm activities as well as in household chores (Fafchamps *et al* 1998, 2003). Therefore the morbidity of adults can lead to labor replacement with child labor. Of course, the death of income-earners in a region with a high AMR may often happen to households, and hence cause many dropouts. In this way, the human capital of the offspring will be depressed and the poverty replicated from generation to generation.

Sawada (1997) also indicated that the effect of transitory income shock on school attendance rate is higher than that of permanent income. Income shock caused by morbidity or even mortality of major income earners of the family is among the most important forms of shock. The findings imply that schemes aimed at helping poor parents to handle emergencies may be more cost effective in enhancing the rate of school attendance than schemes aimed either at reducing poverty itself or reducing school costs for the poor as a whole.

Mortality rate, fertility and rural poverty

Everyone faces health risk, the rural poor face more and the children in the rural poor family face the most. As a consequence, IMR and U5CMR are disproportionately high in most rural areas of the developing countries. The attempt to link the IMR and U5CMR to fertility is made on the basis of the hypothesis of children compensation. Due to the lack of social security in rural area, particularly the pension insurance for the aged, it has been argued that the rural parents compensate for high IMR and U5MR by having more children in order to guarantee a minimal number of surviving offspring (e.g., Nugent 1990; Rosenzweig and Wolpin 1985; Chakrabarti *et al* 1993). This imposes a far-reaching impact on total fertility rate[3] (TFR), population growth rate, crude birth rate, which in turn affects health outcomes, deepens rural poverty and hampers development. For example, declining total fertility rates and birth rates have several important consequences for the health sector. In the short run, greater spacing between births improves the health of both mothers and children. A reduction in actual number of birth reduces the need for obstetric care, immunization and other child health interventions. These often constitute a significant fraction of the public effort on health in low-income countries. In the long run, declining fertility rate may change the age distribution of the population from younger ones toward middle and older ages. The change in age distribution of the population affects the pattern of demand for health services. Also, population growth rate determines the doubling time of the populations.

[3] Total Fertility Rate (TFR) refers to the total number of children a woman would have by the end of her reproductive period if she experienced the currently prevailing age-specific fertility rates throughout her childbearing life. It is expressed as children per woman.

On the other hand, fertility rate has a great impact on a country's population composition. A falling birth rate renders population older (which reduces the proportion of children). A high fertility rate, in contrast, raises population pressure both within household and in society, causing a household or an economy to a Malthusian trap.

The seemingly strong argument has actually some defections. First of all, it is believed that the adult mortality rate instead of IMR and U5CMR ought to be a more suitable indicator for the decision to have more children if people have children to ensure old age support. The reason is that, while adult mortality rate cannot be corrected *ex post*, most infant and child mortality can (Fafchamps1999). However, one indicator can only illustrates health risk in a specific time intervals, it alone cannot show sufficient information to a household. So the combination of IMR, U5CMR with AMR may form the basis of decision on how many kids to have. Both high U5CMR and high AMR induce the risk averse parents concerned with old age support to compensate by having more children, resulting in an average number of surviving offspring than originally anticipated, and thus in an increase in population pressure.

Secondly, in the same manner, MMR and pregnancy-related death are better indicators accounting for high TFR and population pressure. It is reported that the parents prefer male to female kids considering the stronger labor force the males have than females (Cai *et al* 2001). Suppose a couple needs at least one surviving boy offspring but the first three are all girls, needless to say, the couple will decide to have the fourth or even the fifth child because they still need to deduct the chance of dying. The desire for male kids raises total fertility rate and then population pressure. However, this might not have fully explained the fertility behavior of having more children when the first three are girls. Another reason is that girls will have to undertake the obligation to bear children when they grow up, and child-bearing could be quite risky in developing countries as reflected by high pregnancy-related death and maternal mortality rate (MMR). So the future pregnancy-related death of the females helps amplify the effect of having more children than the number the parents planned. The policy of family planning in China, for instance, allowing the rural couples that have already had the first girl baby to have an additional baby after a particular waiting time is partly out of this consideration. A policy implication of the two indicators is that the lower MMR and pregnancy-related death would give households disincentives to have more children and hence drop the total fertility rate.

Working accidents and labor choices

Non-farm work accounts for a varying proportion of income of the rural and is one of the important strategies for coping with crop failure and famine. In Africa, about 60 percent of total household income are derived from off-farm sources. In Gambia, for example, smallholders derive 23 percent of their total income from nonfarm work (von Braun *et al* 1999 and literature therein). Kochar (1999) reports increased labor supply as the key response in the ICCRISAT villages in south India. Working as wage-laborer is even considered a last resort for those lacking assets, access to credits and kin/community support (Sauerborn *et al* 1996a). However, due to bad regulations and monitoring of work conditions in informal sectors, working accidents happen frequently in developing countries, especially in jobs accessible to the rural poor. Coal miners who work underground and builders who work on hazardous building sites while lacking necessary working protection are two examples. On-the-job injuries may have severe consequences: disability and even death. Nevertheless, the employers, usually small entrepreneurs, have not insured the laborers against working accidents. Although they do compensate the disability and death, the amount of compensation is too small to support the dependents of the disabled or dead in the accidents.

Fear of the high incidence of working accidents may preclude people from undertaking these jobs. The rural poor, when faced with an income shock, have to take the risk of engaging in these jobs after running out of their coping strategies. These households that initially hope to pull through the hardship may unfortunately fall deeply into poverty.

2.2 Linkage of health risk to other rural risks

In general, risk is more prevalent in rural areas of the Third World than in developed economies (Fafchamps 1999). Adequate attention has been paid to health risk in the preceding section. Besides health risk, households in rural areas of developing countries are faced with a variety of other risks. Until now the health risk is explored independently. However, the various risks of rural people are interrelated and exposure to one type of risk may lead to occurrence of another risk event. So the distribution of occurrence of different types of risk is auto-correlated over time. Many types of risks are not only intertwined with each other, but they are also coped with jointly with the limited mechanisms (informal risk sharing mechanisms or self-insurance). Thus, it is essential to be aware of other risks in rural areas while maintaining the focus on health risk.

2.2.1 Business risk

Business risk is a rather extensive concept. All risks related to business and resulting in income fluctuation can be referred to as business risk. Unlike most urban residents who earn wages from jobs and then are immune to business risk, almost all the rural households earn their livelihood from the farm, an artisanary workshop, a trading activity, or a services operation. Then business risk of farmers is a very common risk factor. According to the classification of Hardaker *et al* (1997), business risk comprises production risk, which stems from the unpredictable nature of weather and uncertainty about performance of crops and livestock; and price risk, which refers to uncertainty of prices of farm inputs and outputs. Business risk furthermore includes personal risks (for example, the illness or death of people who run the farm) and institutional risk, which originates from the uncertainty of the impact of government policies on farm profits. Now that business is the major source of income, the business risk will naturally lead to fluctuation of income. Sometimes it is also referred to as income risk. Business is rather vulnerable to shocks because the rural poor adopt little advanced technologies and offer little protection against natural events. Moreover, business in poor areas is small, undercapitalized and worse equipped (Daniels 1994).

Business in rural areas may fail for many reasons. In Ethiopia, for example, Dercon (2002) found that 78 percent of the rural households suffered harvest failure, 42 percent policy shock, 40 percent labor problems, 39 percent oxen problems, 35 percent problems with other livestock, 17 percent land related problems, and 16 percent loss of assets in the past 20 years. Harvest failure, the most commonly encountered risk, may result from such climatic factors as flood, drought, frost, hail or insect pest as well as technological failure such as unadapted seeds and way of cultivation. As far as the policy shock is concerned, changes in tax rate and subsidies system might have an immediate impact on income of farmers, while the shift in rural credit system and agricultural product market system exert an indirect influence on income. For farmers, many of production risks belong to collective risk that has an impact on lots of farmers in a region or a community.

Price risk is also predominant, especially in the economy where the market is incomplete and in transition. If the farmers are not familiar with the price mechanism, their incomes will be affected by fluctuation in price tremendously. Fluctuation in price stands for a covariate risk of farmers and is hard to deal with. This affects the rural dwellers in various ways, such as through prices of agricultural inputs and outputs, the value of rural assets, access to off-farm job opportunities, and remittances from migrants (Fafchamps and Gavian 1997).

Although we customarily sort out the risks and let many kinds of risk enter into covariate risk, few risks are purely covariate or idiosyncratic, some studies even found that the idiosyncratic part of income risk is relatively large. Deaton (1997) found that common components explained very little of variation in household income for particular villages in Cote d'Ivoire in 1985-86. Morduch (2002) concluded that idiosyncratic risk (including measurement error) accounted for 75-96 percent of the total variance in income within these villages.

The standard agricultural production function contains only two productive factors: labor and capital. However, the number of labors and capitals cannot interpret all of the production, so the concept of human capital has been notably introduced (Schulz 1961). Human capital can be formed through nutrition, health and education. Through human capital of laborers, health risk of laborers is integrated into business risk. In the process of business risk management, the farmers have to take account of the health risk of laborers or the managers.

2.2.2 Ritual risk

Ritual risk is also an important risk factor in societies, but more a burden for rural residents. In many societies, social customs dictate that customary expenditures should be incurred to mark particular events. Some of these events are fairly predictable; others are not. These customary expenditures serve many purposes, one of which is to rhythm the life of a group and to demonstrate its cohesiveness. The commonly encountered forms of ritual risks involve wedding ceremony, dowry payments, funeral, or even a party for family reunion at seasonal holidays. These ritual activities are sometimes rather costly. Nevertheless, this kind of risk is different from other sources of risk because it is induced by ritual obligations, which result from social pressure more than physical need. It is people's attempt to live up to social expectations and to show their continued memberships to a group that induces them to abide by these social norms. The fairly few evidence suggests that the effort to meet excessive social obligations (e.g., wedding ceremony or funeral) is an important cause of poverty if households borrow money to meet them (Fafchamps and Lund 1999). Even though the ritual risk does not have a direct relation to health risk, the customary expenditure on ritual ceremonies may invade the originally limited physical, human and social capitals and triggers consumption risk, which in turn raises the possibility of malnutrition and health risk. More severely, the limited strategies for coping with risk may have been depleted, leaving the costs of illness uncovered.

2.3 Benefits of risk sharing

Most people are risk averse when faced with significantly risky incomes or wealth outcomes. A person who is risk averse would be willing to forgo some expected return for a reduction in risk; the rate of accepted trade-off depends on how risk averse that individual is (Hardaker *et al* 1997). In practice, people take a variety of strategies to cope with risk and maintain the level of risk at which he or she can bear. Rejda (1997), for example, divided strategies for handling risk into five categories: avoidance, loss control, retention, noninsurance transfers, and insurance. However, according to the criterion of whether the household shares risk with others, there are only two types: non-risk sharing and risk sharing. The former refers to the methods of risk avoidance, loss control, risk retention or non-insurance transfer or some coping strategies such as income diversification, precautionary saving, or borrowing. However, the ability of individual households to cope with risk is quite limited. Therefore, risk sharing (informal or formal) is a more important strategy, whereby many benefits for farmers as well as for the society are generated.

Benefits of risk sharing for the rural households

The benefits of risk sharing could be easily understood from the burden of risk that has been explored previously. In general, the majority of the rural is risk averse and hence is willing to pay a certain amount of premium to reduce exposure to risk. If the rural dwellers can trade away part of risks of business, ritual or health, the utility of the farmer will increase (Arrow 1963, 1965, 1996; Meuwissen 2000). The benefits for rural people from the perspective of microeconomics comprise the following aspects:

- Risk sharing encourages the rural to seek health care or some remedies for production instantly and hence decrease the scale of the possible loss.
- Timely compensation can maintain normal consumptions of rural households and normal productions of the rural businesses.
- Risk sharing may lead to a decrease in risk funds or precautionary savings and hence promotes the amount of investment or investment in high-risk but more-profitable business.
- Risk sharing results in more stable incomes. More stable incomes are likely to lead to more expenditure on farm inputs and family consumptions. More stable consumptions help satisfy rural people's basic needs for nutrition and in turn improve the productivity of laborers and lower the health risk.
- More regular and stable cash inflow enhances a person or a household's credit. So the rural can more easily smooth their income through borrowing from either formal or informal financial institutions or other households.

Benefits of risk sharing for society

There is always the simple justification for any implicitly or explicitly contract freely arrived at between two individuals; if both of them choose to enter the contract, then both of them must be better off. Society after all is simply a convenient label for the totality of individuals and if the two contracting parties are better off, so is society, unless other individuals are injured in some way. The existence of risk sharing also results in great benefits to society as a whole.

- The possibility of sharing risk permits individuals to engage in risky but more-profitable activities, which they would not undertake otherwise. By this means, the expected return to society increases over what would prevail if individual agents were constrained to accept only those risks they could afford to bear themselves (Arrow 1965; Hardaker *et al* 1997; Rejda 1998).
- If farmers need to put less effort into on-farm methods of avoiding risks, they will be able to use their resource more efficiently, which implies greater overall efficiency in resource allocation (Hardaker *et al* 1997; Rejda 1998).
- Stable income should be reflected in improved access to credit and/or lowered borrowing costs, implying increased productive investment in poverty reduction.
- If the rural poor are able to trade away part of their catastrophic risk, the

resilience (or sustainability) of the farms increases, which may mean less human, animal and environmental distress after the occurrence of disaster such as flood or droughts. This, however, is only true if moral hazard is dealt with properly. Otherwise farmers may, for example, pay less attention to the prevention of disease, increasing the incidence of disasters.

- Technological innovations have already been proved to be the ultimate engine of economic growth theoretically and empirically. However, the production of new and advanced technology is a venture into the unknown. The outcome of any research project is necessarily uncertain, and the most important results are likely to come from projects whose degree of uncertainty to begin with was greatest. The risk sharing of scientific research and technological innovation is thus most needed from society's point of view.

2.4 Who provides health insurance for rural people?

Health insurance is a modern institution of risk sharing, which pools and spreads health risks in a huge amount of people so that the probability of risk event is certain. Basically health insurance can be provided privately. Nevertheless, there are failures in the private health insurance market. An unregulated private insurance market suffers from adverse selection and moral hazard by consumers, and risk selection by insurance companies. Moreover, insurance companies have demonstrated a strong propensity to create monopolies or cartels. The market failure results in no insurance coverage for the poor, the aged, and the disabled and yield excess profit for insurance company (Hsiao 1995). Therefore, the government intervention in the health sector becomes required. Another rationale for government intervention in the health sector is market failure in health care, which is in possession of the attributes of externalities, public goods, and merit goods. Basic health care is not a public good, for the consumption of health care can exclude others (Folland *et al* 1997). But basic health care is associated with large positive externalities, that is, marginal social benefit exceeds marginal social cost, leading to the sub-optimal production and higher-than-optimum prices. Health insurance helps mobilize more resources and revenues for health care. Therefore, theory suggests that government intervention in health insurance market possibly improves welfare if government has better information than individual market participants (Arnott and Stiglitz 1990; Arnott 1990).

However, the government intervention is not a panacea, even in provision of public goods. It is common that some goods are overprovided for some group of people in some areas while they are underprovided for others in other areas. The purpose of political leaders to maximize the likelihood of staying in office occasionally contradicts the hope of optimal provision of public goods for the public. In the case of health insurance, the urban residents who are working in formal sectors and are politically powerful readily enjoy the social health insurance on the basis of employment. In contrast, the rural population who are less influential in politics is institutionally difficult to be covered by social health insurance scheme. Government intervention is generally assumed to have information advantage over the private sector. However, that is not the case in health insurance market. The government does not know more about the prospective insured than private insurer to avoid the problems of adverse selection and moral hazard, which are believed to lead to market failure of health insurance. So government failure also exists.

The community, a group of people tied by mutual trust based on intense personal interaction, can provide organizations critically needed to correct the failures of market and the state (Hayami 1997). Generally speaking, the community has comparative advantage over state and market in supplying local public goods. In the case of health insurance, the incidence of moral hazard should be lower among people whose personal interactions are intense so that each can predict accurately the other's behavior. Also, mutual trust nurtured through close personal relations should work as an effective brake on committing moral hazard. An institutionally innovative arrangement to cope with health risk in rural community that has been recently re-mentioned is community-based health insurance scheme (CBHIS). Empirical studies have provided us with many examples of CBHIS in different forms in different countries. The community involvement in user fee collection is exemplified with Bamako initiative in Benin and Guinea; Grameen in Bangladesh, mutual health organization in Thies and Senegal as well as RCMS in China are examples of community prepayment scheme or mutual health organization. In addition, Nkoranza in Ghana belongs to provider based community health insurance and Thai health card scheme is a case of community driven pre-payment scheme attached to social insurance or government run system (Jakab *et al* 2001). A hypothesis that both the active participation of community and the deep involvement of government underlie the successful community-based health insurance seems to have been confirmed and will be further tested in the study.

In spite of the advantage in dealing with adverse selection and moral hazard, the community's ability to cope with the covariate risk is fairly limited and the objective of health equity for all can hardly be reached owing to the wide discrepancy in level of income between communities. Therefore, while we strengthen the effectiveness and efficiency of community based health insurance scheme, a suitable degree of involvement of government should never be disregarded.

The researcher argues that there is a big role for market, government and community to play in health insurance provision. Correspondingly, a combination of a variety of forms of health care financing mechanisms, whether commercial health insurance, community-based health insurance scheme or even informal insurance arrangements like the extended family, can strengthen the ability of the rural population to cope with health risk and provide them with a good protection against costs of illness.

3. An Overview of China's Health Care System

3.1 The country

3.1.1 Physical features

Located in East Asia, on the west shore of the Pacific Ocean, the People's Republic of China has a territory area of 9.6 million square kilometers as well as maritime area of 4.73 square kilometers, and is the third largest country in the world, next to Russia and Canada. China has land borders 22,800 kilometers long, with 15 contiguous countries: Korea to the east; Mongolia to the north; Russia to the northeast; Kazakhstan, Kirghizstan and Tajikistan to northwest; Afghanistan, Pakistan, India, Nepal, Sikkim and Bhutan to the west and southwest; and Vietnam, Laos and Myanmar to the south. Across the seas to the east and southeast are South Korea, Japan, the Philippines, and Brunei, Malaysia and Indonesia.

China's administrative units are based on a three-level system dividing the nation into provinces, counties, and townships. At present, China is divided into 23 provinces, 5 autonomous regions, 4 municipalities directly under the Central Government, and 2 special administrative zones (Hong Kong and Macao).

China's topography is varied and complicated, with towering mountains, basins of different sizes, undulating plateaus and hills, and flat and fertile plains. A bird's-eye view of China indicates that China's terrain descends in four steps from west to east. China is also abundant in rivers. More than 1500 rivers each drain 1000 square kilometers or larger areas. The catchment's area for the rivers that empty into the oceans accounts for 64 percent of the country's total land area. The Yangtze, Yellow, Heilongjiang, Pearl, Liaohe, Haihe, Huaihe and Lancang rivers flow eastwards, and empty into the Pacific Ocean. The Yarlungzangbo River in Tibet, which flows eastwards first and then southwards into the Indian Ocean.

China has a marked continental monsoon climate. The four seasons are quite distinct. The rainy season coincides with the hot season. The complex and varied climate results in a great variety of temperature belts, and dry and moist zones. In terms of temperature, the nation can be sectored from south to north into equatorial, tropical, sub-tropical, warm-temperate, temperate, and cold-temperate zones; in terms of moisture, it can be sectored from southeast to northwest into humid (32% of land area), semi-humid (15%), semi-arid (22%) and arid zones (31%).

The composition and distribution of China's land resources have three major characteristics: (1) variety in type—cultivated land, forests, grasslands, deserts and tideland; (2) more mountains and plateaus than plains and basins; (3) unbalanced distribution: farmland mainly concentrated in the east, grasslands largely in the west and north, and forests mostly in the far northeast and southwest. 94.97 million hectare of land cultivated is distributed mainly in the

Northeast Plain, the North China Plain, the Middle-Lower Yangtze Plain, the Pearl River Delta Plain and the Sichuan Basin. In terms of sheer area China's cultivated lands, forests and grasslands are among the world's largest. However, per capita areas of cultivated land, forest and grassland are quite small due to its large population, especially with respect to cultivated land—less than 0.08 hectare per capita, namely, a third of the world's average.

3.1.2 Population and family in China

According to the communiqué of the Fifth National Census at the end of 2000, China, the most populous country in the world, has a population of 1.295 billion, accounting for some 22 percent of the world's total. This figure also includes Chinese in the Hong Kong Special Administrative Region, Taiwan Province and Macao Special Administrative Region. The population density in China is 130 people per square kilometer. This population, however, is not evenly distributed. China has 56 nationalities, the Han is majority and accounts for 91.6 percent of the total population, with the remaining 8.4 percent being minorities.

Since the end of the 1960s, China has adopted a basic policy of *family planning* to control the fast growth of population. The family planning basically requires people to marry late, give birth to baby late, and requires one couple to have one child. As a result, the type of population reproduction turns into "low birth rate, low mortality rate and low natural growth rate"(Cai and Dou 2001); annual average growth rate in the past decade is 1.0 percent (World Bank 2002). The family planning had also led to the drop in household size, from 3.96 in 1990 to 3.44 persons in 2000. As a consequence of low natural growth rate, the society began aging; people above 65 years old increased by 1.4 percent and reached up to 7.0 percent at the end of 2000. People become better educated. The illiterate rate fell by 9.2 percent and is now 6.7 percent. Along with the industrialization and urbanization, the rural population takes up 64 percent. Notwithstanding the intended outcome of family planning, it remains a formidable task to achieve the goal of population development: to control the population within 1.33 billion in 2005 and within 1.4 billion in 2010 (not including those in Hong Kong SAR, Macao SAR and Taiwan province); to achieve zero population growth in the middle twenty-first century.

China has 332.97 million families, with 3.2 persons per urban household, and 4.3 persons per rural household on average. In general, a Chinese family is composed of a couple and their children, but big families with three or more generations can also be found in China. Along with the pursuit of personal freedom, the trend of forming small families with only directly related members is now prevalent. In the past, each Chinese family had a head, who had absolute authority at home, and had the final say in family affairs. But now in most Chinese families, the husband and wife, or a couple with other family members, work out the household plans together, and decide family affairs through consultation. Moreover, family members share the housework, making the division of labor at home more reasonable; and the husband and wife support each other's work. China, one of the world's most ancient civilizations, has a recorded history of nearly 4 000 years. In the history, there were many schools of philosophic thinking and ideologies. Finally, the Confucianism has been dominated in stipulating the behaviors of common people. On the basis of the Confusianism, the tradition of respecting the old and loving the young was formed and comes down. Though many young couples do not live with their parents, they maintain close contact with them. Grown-up offspring have the duty to support and help their parents. The Chinese also attach great importance to relations between family members and relatives, and cherish their parents, children, brothers and sisters, uncles, aunts and other relatives. Additionally, people put emphasis on neighbors and friends. Mutual help is always regarded as a virtue in Chinese culture.

3.1.3 Socio-economic development

Since 1949, despite the political consultation with other democratic parties, the Communist Party of China (CPC) has de facto been ruling the People's Republic of China. After a long period of political unrest, the Party finally recognized that the most important task should be to prompt economic growth by reform and opening door to outside. According to Deng Xiaoping's theory that is now the guiding ideology of China, the root task of CPC is to build socialism with Chinese characteristics to guarantee the sustained, rapid and healthy growth of the national economy, and realize the modernization in the middle 21st century.

To fulfill this root task, China launched economic reforms in a gradual way. The pilot reform took place in rural areas in 1978 by transforming the agricultural production system from a collective one to household contract responsibility system. It soon proved to be highly successful and agricultural output rose quickly because of the enhanced incentives of farmers. Inspired by success in the reform of agricultural production system, the government launched urban reform subsequently. In the period before 1992, the *centrally planned economy* had evolved into an alleged *planned commodity economy* and hence the commodity market and factor markets emerged and developed gradually. In 1992, the *market-oriented economic system* was definitely put forward and began to be utilized as the main channel of allocating resources. The whole country was stimulated by the new mechanism and almost all sectors, enterprises and workers got new impetus to economic growth, national economy not only develops in sustainable, rapid and healthy pattern, but also the economic structure become modified. The role of agricultural sector has been lessening and the industry contributes more to the Gross Domestic Production (GDP), especially the output of manufacturing got a stable increase. A series of indicators of economic performance and Human Development Index[4] (HDI)

[4] HDI is a composite index measuring average achievement in three basic dimensions of

can manifest the social economic development.

As reported by World Bank (2002) in Table 3-1, the economic reform brought about the economic growth at average annual rate of 10 percent from 1981 to 1991 and 9.7 percent from 1991 to 2001. The GDP has reached US$ 1 150.1 billion in 2001. Gross National Income (GNI) per capita is US$ 890, which is close to the average level of East Asia and Pacific Countries and higher than the average GNI per capita of developing countries as a whole.

Table 3-1 Key indicators of macroeconomic performance of China

Indicators	1981	1991	2001
GDP (US$ billion)	228.3	402.6	1150.1
% of GDP			
Agriculture	31.8	24.5	15.2
Industry	46.4	42.1	51.1
Manufacturing	38.3	32.7	35.1
Services	21.8	33.4	33.6
Gross domestic investment/GDP	32.5	34.8	37.9
Gross domestic saving/GDP	32.9	38.1	25.8
Total exports (US$ million, fob)	22,007	71,843	266,155
Total imports (US$ million, cif)	22,015	63,791	243,610
Current account balance (US$, million)	860	15,240	17,405
Reserves including gold (US$ million)	-	48,154	219,970
Foreign direct investment (US$ million)	-	4,366	47,052
Government finance: Overall surplus/deficit (% of GDP)	0.8	-1.1	-3.2
Registered unemployment rate (%)			3.6
Gini Coefficient			0.45

Source: World Bank 2002; NSB 2002.

The fast economic growth produced favorable welfare improvement for people to a large extent. China's HDI increased from 0.554 in 1980 to 0.726 in 2000. In terms of education, more people can read and write than in 1980 and the literate rate is even higher than that of the developing world as a whole. Furthermore, Chinese enjoy a longer and healthier life now than in 1980. The life expectancy and GDP per capita in terms of purchasing power parity (PPP) are higher than the average level of developing countries (Table 3-2). Both health and education indices are higher than the GDP index, revealing that people in China enjoy a relatively good welfare compared to the level of economic development.

China has significantly reduced rural poverty over the past few years. Sustained, rapid economic growth and targeted poverty intervention brought the number of the absolute rural poor in terms of the official poverty line down from 80 million in 1994 to below 30 million, or about 3% of the rural population, in 2001. However, the official poverty line was set in the mid-1980s and is

human development-a long and healthy life, knowledge and a decent standard of living. It is also a quite authoritative indicator calculated by the UNDP and published in yearly Human Development Report.

substantially lower than the international standard, one dollar per day. About 213 million people, or 23% of the rural population, still live below that level (ADB 2002). There is still a long way to go for combating against poverty.

Table 3-2 Human Development Index in China and other groups in 2000

Indicators	**China**	Developing countries	East Asia and the Pacific	OECD
GDP per capita (PPP$)	**3 976**	3 783	4 290	23 569
Life expectancy at birth (years)	**70.5**	64.7	69.5	76.8
Adult literacy rate (% age 15 and above)	**84.1**	73.7	85.9	..
Combined primary, secondary and tertiary gross	**73**	61	71	87
Life expectancy index	**0.76**	0.66	0.74	0.86
Education index	**0.80**	0.69	0.81	0.94
GDP index	**0.61**	0.61	0.63	0.91
Human development index (HDI) value	**0.726**	0.654	0.726	0.905

Source: UNDP 2002.

In addition, some serious problems arose in China. After over two decades of economic reforms, China's dual economy, the coexistence of developed urban industry and backward rural agriculture, is continuing. About 64 percent of population still lives mainly on income from agricultural production in rural areas. The gap of income between urban and rural areas, between the eastern and western regions is widening, with an alarming Gini coefficient of 0.45. The government still finds no way out of the difficulty in reforming state-owned enterprises that caused high registered unemployment rate in cities as well as a large amount of laid-off workers; The heavy burden of government finance and high ratio of bad financial assets are slowing down many further reforms.

3.2 The health sector in transition

The issues on the health sector could be better addressed if the urban and rural sectors are treated separately rather than mixed[5]. After all, nearly 64 percent of population still lives in rural areas. New migrants who have resettled in the towns or cities seem unlikely to participate in urban employment-based health insurance and hence enjoy equal access to health care.

3.2.1 Health care delivery system and physical accessibility

China's health care delivery system has been documented in some of past studies (Dong 2000; Wu 1997; Ho 1994). Along with the reforms of the health

[5] Grogan, Collen M. (1995) believes that the process of focusing on health services in rural and urban areas as separate systems is unnecessary and even counter-productive in light of the emerging integration of China's urban and rural-economies.

sector, however, the delivery system has been in great mobility. The review of the recent change may produce a better understanding of the reforms. It is usually thought that health administration is health bureaus at all levels, from the Ministry of Health to county health office. As seen in the Table 3-3, the health administration has four tiers in spite of China's five-level of governments; health care in township and village is administered and guided by county health bureau. As a matter of fact, the Ministry of Labor and Social Security and its all levels of Bureaus are also deeply involved in the health sector, but its focus is placed on health care financing. In addition, the policy of Ministry of Agriculture is related to the administration of rural health care delivery and financing system.

Table 3-3 Health administration and health care delivery system in China

Health administrative institution (Top down)	Health care facilities (top-down)	Target population
Ministry of Health, National Bureau of Medicinal Administration and Monitoring (newly founded)	Key medical teaching hospitals; Medical academies; Others	
Provincial Health Bureau	Provincial hospitals Provincial preventive stations Provincial health centers for women and children Hospitals affiliated to local medical university	Urban residents in provincial capital city; The severe sick referred from lower level facilities
Prefecture Health Bureau	Prefecture hospitals; Prefecture preventive stations Prefecture health centers for women and children	Urban residents in prefecture cities; The severe sick referred from lower level health facilities
County Health Bureau	County hospitals; County preventive stations County health centers for women and children	Urban residents in county towns; Suburban rich farmers; The severe sick referred from lower level facilities
	Township health centers;	Both rural and urban residents in the township.
	Village health stations	Local villagers

Source: own compilation

To satisfy the need for health care of both the urban and the rural residents, an elaborately designed three-tier delivery system was established. For the urban residents who reside intensively in cities, the urban three-tier system originally comprised street health stations, community health centers and district hospitals. However, the urban three-tier system broke down due to the competition mechanism introduced into the health sector. The primary and secondary tiers have been merged into the district hospital or turned into district clinic. Health care can nowadays be delivered to the urban by health facilities at the level of county, prefecture, province as well as health facilities affiliated to state-owned enterprises, universities and army.

It is far more difficult to deliver health care to the rural dwellers because they reside in a quite scattering way. Health care has also been delivered via a rural three-tier delivery system, which consists of village health stations, town health centers, and county hospitals. Scales and basic conditions of three-tier health care delivery system and private clinics in China are presented in Table 3-4. Village health station is the most basic tier and has been established on an administrative village basis. Village health station is, according to the statistics end of 2001, on average staffed with 1.8 village doctors who are trained for three to six months after junior high school and are received on average two to three weeks of continuing professional education each year. They provide merely outpatient care for some commonly encountered diseases, maternal and child care, and some preventive care. This level of facilities used to be the first contact places where rural patients sought medical care. The patients would be referred to upper tier, township health center, in case the disease is beyond their capacity of treatment. There are totally 700,000 village health stations in China and nearly 90 percent villages each have one health station.

Table 3-4 Three-tier health care delivery system & private clinic in rural China

Types of health institutions	Number of health institution	Total hospital beds	Average hospital beds	Occupancy rate	Average hospitalization days	Total number of staff	Average number of staff
County hospital	3 676	462 236	126	0.613	14.2	663 097	180.4
Town health center	48 090	740 060	15	0.313	4.5	1 168 932	24.3
Village health station	698 966	0	0	0	0	1290595	1.8
Private clinic	136 561	n.a	n.a	n.a	n.a	194 260	1.4

Source: cited or calculated from statistic data of MOH 2002

Town health centers, the second tier of the system, have on the average 15 beds overseen by 24 health workers, who are composed of physicians with three-year medical school education after high school, assistant physicians and village doctors (NSB 2002). They provide both outpatient and inpatient services depending on the equipments and technological ability of the health workers. The common services range from treatment of common diseases, simple abdominal operations, X-rays and laboratory tests, maternal and child health care as well as preventive services. They also provide the village doctors with some technical guidance. There are in total about 48,000 town health centers nationwide, every town (or township) has a minimum of one health center.

The tertiary tier comprises the health facilities at the level of county. There are usually at least two county hospitals in each county, a western medicine hospital and a Chinese medicine hospital. There are also several preventive facilities: preventive station and health centers for women and children. On average, each hospital usually has 126 beds and is staffed with 180 health workers, including physicians of four to five years of college education as well as nurses and technicians and pharmacists. The county hospitals provide almost all types of out- and inpatient services, operations, tests as well as western and Chinese medicines.

In general, health care is geographically accessible to Chinese thanks to the three-tier health care delivery system. As shown in Table 3-5, 70.7 per cent of homes are below 1 km away from the closest health care facility and 68.8 percent of people can arrive at the closest health facility within 10 minutes by means of public transport or on foot. However, the physical accessibility is less satisfied in rural areas. Some 4 percent of homes are as far as over 5 kilometers and 7.7 percent of population has to travel over 30 minutes to reach the closest health facility. Moreover, a great many of the closest health facilities are village health stations, which provide insufficient and unsatisfied services.

Table 3-5 Physical accessibility to health care in China in 1998

Indicators	Sum	Urban areas	Rural areas
Distance from the closest health facility (%)			
Less than 1 km	70.7	77.5	67.9
1-5 km	26.1	21.7	27.9
More than 5 km	3.2	0.8	4.2
Time spent traveling to closest facility (%)			
Less than 10 minutes	68.8	72.4	67.4
10-30 minutes	25.2	25.7	24.9
More than 30 minutes	6.0	1.9	7.7

Source: MOH 1999

This three-tiered system was aimed to promote the efficient allocation of health care resources between primary and tertiary care facilities. Traditionally, the three-tier system was indeed for several decades an efficient patient-referral system for health problems to be treated in the most appropriate setting. The rapid changes in society and economy, however, have brought new challenges to this system. With the growth of income and improvement of transportation, a portion of patients are willing to pay more to go direct to the county hospitals. The responsiveness to quality of care and income become high and there is significant shift in demand for health care from village health station to upper tiers (Hsiao 1994). Many of the village health stations, therefore, find it difficult to maintain operation, especially in the wealthy and transportation-developed villages. In addition, the development and reform of the health care delivery system is intimately connected with that of the health care financing system: the two systems are complementary and mutually restricting. Next, we turn to the health financing system.

3.2.2 Health care expenditure and health care financing system

3.2.2.1 Health care expenditure

In the past decade the health input increased sharply. According to the latest statistics shown in Table 3-6, the total nominal health expenditure rose at a rate of two digits and amounted to 476.4 billion Yuan by 2000. In terms of weight in GDP, health input escalated from 4.1 percent in 1990 to 5.3 percent in 2000 and in terms of health expenditure per capita it from 76.7 to 376.4 Yuan. Three main facts can be summarized from Table 3-6. First, the total health expenditures increase faster than GDP. Second, the government and society contribute a decreasing share and people become more reliable on their own income to pay the health bill. Third, the urban residents spend much more in health care than their rural counterparts. As a result, the large amount of health input actually is not uniformly distributed across population groups. A handful of studies showed the health inequality between the urban and the rural residents (Liu *et al* 1999; Gao and Qian *et al* 2002). To deliver health resources to specific individual or population group depends to a large extent on the health financing system.

Table 3-6 Health expenditures in China 1990 to 2000 **Units: Yuan or % for rate**

Year	GDP		Total health expenditures			Structure of total health expenditure (%)			Health expenditure per capita		
	Nominal (billion)	Growth rate	Nominal (billion)	Growth rate	Weight in GDP	Revenue	Social	Private	Urban	Rural	Sum
1990	1859.8	3.8	138.5	14.3	4.1	25.0	38.0	37.1	158.8	38.8	76.7
1995	5749.5	9.0	235.2	12.8	3.9	17.0	32.8	50.3	401.3	112.9	186.4
1996	6685.1	9.8	281.0	19.5	4.2	16.1	29.6	54.3	467.4	150.8	233.5
1997	7314.3	8.6	330.3	17.6	4.6	15.4	27.7	56.9	537.9	177.9	273.8
1998	7696.7	7.8	377.7	14.3	4.8	15.6	26.6	57.8	595.3	193.9	302.6
1999	8042.3	7.1	417.9	10.7	5.1	15.3	25.5	59.2	-	-	331.9
2000	8940.4	7.9	476.4	14.0	5.3	14.9	24.5	60.6	-	-	376.4

Source: NSB 2001 and 2002; MOH 2001

3.2.2.2 Structure of health care financing in China

Health care financing system mainly refers to the economic mechanism regarding the collection, distribution and utilization of health resources (Gu 1995). The effective provision of health care, apart from a well-structured health care delivery system, depends on health resources available in a country. Health care financing system determines to whom the resources are allocated and to what extent. Traditionally, China's health care financing system contains three pillar health insurance schemes: government employee health insurance scheme (GHIS) and labor health insurance scheme (LHIS) in urban areas, and rural cooperative medical scheme (RCMS). In addition, laborer's dependants health insurance scheme (semi-LHIS), the private health insurance, and out-of-pocket payment are alternatives. However, urban employees and workers are joining the same social insurance—the basic health insurance of urban employee—after an overall reform in urban health care financing system starting in 1999.

The health care financing system for China's urban workers was established in the early 1950s. Financed by government budgets, government health insurance scheme (GHIS) covers government employees, retirees, disabled veterans, and university teachers and students. Labor health insurance schemes (LHIS) were required to establish in state-owned enterprise (SOE) with more than 100 employees and cover their employees, retirees and their dependants. Smaller state enterprises and village & township-owned enterprises can provide LHIS on a voluntary basis. In 1998, GHIS and LHIS covered approximately 4.95 and 6.22 percent of the total population, respectively (MOH 1999).

In suburban and rural villages, the predominant health insurance scheme has been the *rural cooperative medical scheme*[6] (RCMS), which was organized at the brigade or village level with funds generated by the villages themselves before the economic reforms. Typically, RCMS generated its funds from annual premium paid by individuals and annual appropriations from village and township collective welfare funds. The premium was based on community rating so that all members, irrespective of their age, sex, education level or current health status, paid a flat premium. The level of premium was determined by the community's health expenditures in the previous year, availability of village fund and enrollment rate. Farmer's contribution ranged from 0.5 to 2 percent of their annual net income. This system enabled farmers to have the access to primary and preventive health care. By the mid-1970s, more than 90 percent of rural populations had joined in RCMS (Gu *et al* 1995; Kahn 1997; Carrin 1999).

It is widely believed that the market-oriented economic reforms caused the breakdown of RCMS. However, the market-oriented reforms prompted the substantial development of the private health insurance (PHI), especially in past five years. The private health insurance policies have been booming. The latest official figure of PHI coverage was only 1.9 percent (MOH 1998). The premium income of health insurance increased at the rate of 27 percent from 1998 to 2000, much faster than aggregate premium income does, according to the report of China's research team on private health insurance headed by Wu *et al* (2002). It is also estimated that 38 million people, or 2.6 percent of the total population purchased health insurance policies by the end of 2000. More than half of the insured are students and children. It is expected that the purchase rate of the rural residents was not lower than that of their urban counterparts in the past few years despite their lower income, since the majority of rural residents cannot rely on the social health insurance and cooperative medical schemes.

The latest statistics from the China's Ministry of Health (MOH) shows that the enrolment rate of rural cooperative medical scheme remains quite low after

---

[6] It refers not only to village or town-based health insurance scheme but also to rural three-tier health delivery system: village health station, township health center and county hospital and their administrative and technological mechanism. For convenience, RCMS refers to only the insurance scheme in the study if it is not particularly pointed out.

twenty-year economic reform. As shown in Table 3-7, only 12.6 percent of the rural residents were covered by some of the insurance schemes while some 56 percent of the urban were insured in 1998. The inequality in health care financing between the urban and rural areas and even within the rural areas is obvious.

Historically the percentage of rural Chinese covered by RCMS was not that low, which started in 1955 with coverage of 10 percent of rural population and then reached the highest in mid-1970s when 90 percent of the rural population was covered. After the economic reforms in 1978 RCMS collapsed and the present coverage is less than 10 percent (Khan *et al* 1997). A time series of data of coverage of the dominant health insurance in rural areas, RCMS, is plotted in Figure 5-1, which shows the rise and fall of RCMS over the period 1958 to 1993. The rapid decline of RCMS since the early 1980s has been well documented and widely analyzed in the literature (Khan *et al* 1997 and literature cited therein; Chen *et al* 2001).

Table 3-7 Structure of health care financing in China in 1998 Unit: %

	National	Urban area	Rural area				
Health care financing forms	Total	Total	**Total**	1st type	2nd type	3rd type	4th type
GHIS	4.91	16.01	**1.16**	1.07	0.76	1.98	0.26
LHIS	6.16	22.91	**0.51**	1.40	0.52	0.15	0.03
Semi-LHIS*	1.60	5.78	**0.19**	0.64	0.10	0.07	0.05
PHI	1.87	3.27	**1.39**	2.39	1.57	1.16	0.12
Risk pooling**	0.39	1.42	**0.04**	0.15	0.03	0.01	0.00
RCMS	5.61	2.74	**6.57**	22.22	3.61	1.61	1.83
Out-of-pocket	76.42	44.13	**87.32**	71.79	92.53	94.78	81.49
Others	3.04	3.73	**2.81**	0.34	0.88	0.23	16.22

Source: MOH 1999

Note: * Dependents' insurance for children of those with laborer insurance; ** Special health insurance schemes run by state-owned work units; ***The rural are divided into four groups according to the level of economic development: first group refers to the richest quartile and the fourth the poorest quartile.

3.2.2.3 Supply of health insurance in rural China

Private health insurance products and supply characteristics

There are three major private insurance companies that supply private health insurance in rural areas: China Life Insurance Company (CLIC), Ping'An Insurance Company of China and China Pacific Insurance Corporation. The focus is then put on CLIC for it accounts for some 68 percent of the private health insurance market in terms of the number of insurants (Wu *et al* 2002) and provides the rural inhabitants with extensive health insurance services, whereas the two other insurers play a relatively small role in insurance provision for the rural, mainly for the rural living in the suburb. CLIC is a state-funded national life insurance company, subjected to the State Council's leadership, which was separated from the former People's Insurance Company of China (PICC) and derived itself from the Zhongbao Life Insurance Co. Ltd. founded in February 1996. It was registered as the current name in January 1999. The company mainly deals in life insurance, which uses life and body as object. There are presently more than 80 insurance products belonging to 3 categories: ordinary life insurance, health insurance and accident injury insurance. So far 0.4 billion person-times have already been insured by the Company. It is one of the most powerful companies in China with complete organization network in both urban and rural areas, there are in total 3400 branch organizations and 20,000 agencies, and 300,000 pluralistic agents.

Health insurance products can be categorized into two: individual health insurance and group health insurance. However, for purpose of convenience, based on the definition of health insurance in broad sense, health insurance in this study refers to coverage's of outpatient and inpatient medical costs and living allowances in hospital due both to disease and to accident injury. All the existing health insurance products on sale by 2001 are presented in Table A-1 of Appendix 1 and some of them sold in rural areas are marked with a star.

Some supply characteristics of private health insurance can be summarized as follows. First, only dread diseases are covered. Major individual insurances fit for the rural dwellers cover some of dread diseases only. For example, "dread disease benefit" covers only 8 kinds of dread diseases or surgeries that many people believe it is almost impossible to encounter: myocardial infarction, bypass of coronary artery, cerebral apoplexy, uremia, cancer, paralysis, major organ transplant, severe burn, fulminant hepatitis and surgery of main artery. A lot of common diseases or a few dread diseases are excluded from the coverage. "Green shade for life—major diseases protector" is an exception covering 581 kinds of diseases. However, it sets both overall and per cause maximum benefit. Thus, the demand for these insurance policies is negatively affected.

Second, most types of health insurance policies set a maximum claim, the maximum amount of money that the insurer is obligatory to pay. Moreover, the quota is dual, both during insurance period and per cause and it is normally low.

Third, town health center is not designated as eligible health facilities to provide health care for the insured due to the disordered management and incapability of self-discipline in running. In some places, the rural households are located in a quite scattered way. The geographic barriers may hold back the remote villagers from seeking care, and consequently from purchasing health insurance policies. However, there are some exceptional cases. Taking example for the "pupil medical and accident insurance", some emergent diseases or accident injuries treated at nearby town health center are also permitted.

Finally, the private health insurance policies lay stress on the function of savings while somehow neglecting the function of security. The reason underlying the phenomenon is the Chinese tradition of savings. The China's insurance industry started much later than the banking and has been growing during the competition with the banks. The insurers succeeded in attracting buyers through the insurance products with the functions of both savings and security.

Rural cooperative medical scheme and supply characteristics

On the basis of the degree of risk sharing, rural cooperative medical schemes (RCMS) can be classified into 4 types in China: welfare type, risk type, risk-welfare type, and comprehensive community health insurance. Each type has different objective, technical and managerial feature and is applicable to rural areas with different level of income in Table A-2 of Appendix A.

Internationally there are four often-encountered forms of community-based health care financing system in terms of the technical design, management, organization and institution characteristics (Jakab *et al* 2001). It is widely recognized that China's RCMS belongs to the third type—community prepayment scheme or mutual organization. Some unique features emerging recently are noteworthy.

The principle of technical design is fixing the benefit on the basis of premium income, keeping income and expenditures balanced. The premium is normally prepaid and both idiosyncratic and covariate types risks are shared within village or town, either outpatient or inpatient services or both are covered depending on income level as well as the preference of local residents; exemption from premium is usually possible for the most poor, e.g., the childless aged. There is no waiting period and the residents join RCMS voluntarily, but the severe ill is excluded.

Usually, the town governments play a major role in initiating, guiding and administrating RCMS while the villages and county governments assume a subsidiary role. One innovative institutional arrangement is the commission of RCMS composed of the director of town health centers, some town officials concerned and representatives the farmers, serving as the principal decision maker and monitor. The commission of RCMS sets up an office in town health center mainly dealing with the reimbursement and preventive health care provision. The director of town health center is normally the executive of the commission of RCMS. The financial affairs are published 3 to 5 times per year. The commissions of RCMS designate town health centers as the major health care providers. The patients can also be referred to county hospitals with the consents of town health centers, but the co-payment rate will become a bit higher.

Prepayment schemes

In 1994 the national congress published *Maternity and Baby's Health Care Act*, which came into effect in the following year. Correspondingly, health administration of provinces (1994) constituted two concrete and exercisable regulations: "detailed administrative rules of dispersedly-resided children's health care system" and "administrative measures of rural pregnant and lying-in women's systematic health care". Meanwhile, another law, *Preventing Infectious Diseases Act,* entitled children to essential immunization. Based on these laws and regulations, the country health bureaus provide three prepayment schemes: prepayment schemes for immunization, comprehensive prenatal care, and maternal & well-baby care (Table A-3 in Appendix A).

These schemes played an important role in ensuring target population to receive timely, regular and essential health care, whether they are well off or in poverty and hence helped the rural residents improve the overall health status. These schemes function mainly as savings system, but also include an insurance element. For example, the pregnant women at high risk or babies of abnormal symptoms may receive more health check-ups or cares than normal ones and do not need to pay for the extra care. Moreover, the health facilities can benefit from the prepaid funds through full preparation of drugs and equipments. The main purpose, however, is to stimulate the demand for preventive services. They are usually organized at county or township level and sponsored by the local government; contracts are made with village doctors to provide services (Bogg *et al* 1996).

Despite the guarantee by law and provincial regulations, the situation of scheme's operation is far from optimistic. These prepayment schemes actually broke down in many counties, where maternal and baby health care or immunization services are provided on the basis of fee-for-service. The health administration and the health facilities do not have strong incentive to promote these schemes because in many cases the price of care in the way of prepayment is actually much lower than the price in the way of fee-for services, the former equals merely 50-70 percent of the latter. It is reported that in developed provinces such as Guangdong and Guangxi the participation rate of prepayment scheme for immunization is respectively 42.9-66.9 percent in 1998 and 46.7 percent in 1999 (Zhong *et al* 2002).

3.2.3 Health status and health risk

"*If you just go through the figures – life expectancy, infant mortality, and the first three causes of death – without knowing which country it is, it would be very difficult to pick the country as China and almost impossible to pick it as a developing country. The top causes of mortality in the cities are the same as in developed countries*", said Dr. B. P. Kean, director of Child and Adolescent Health in WHO. Since the establishment of People's Republic of China the health status has grown substantially. In mid-1950s most people were in very poor health due to the bad nutrition and shortage of health care. Infant mortality rate (IMR) was 138.5 per thousand live births in 1954 and the life expectancy (LE) was 57 years old in 1957. The IMR decreased to 34.7 per thousand in 1981 and the life expectancy increased up to 67.9 years (MOH 1999). The latest statistics exhibits that Chinese lead a longer life up to 71 years while 32 new births per thousand die from various types of diseases (MOH 2002).

The national health status indicators show that China has achieved a very good health outcome with its limited health inputs. The performance is quite outstanding when compared with that of the whole developing world as seen in Table 3-8. This outstanding health outputs have been obtained with only half the inputs of the average level of the developing world as a whole. The performance is also better than that of East Asia and Pacific countries.

Table 3-8 Selected health indicators of China and other economies

Health indicators	China	Developing countries	East Asia & Pacific	Germany
Health expenditure per capita (US$) (1999/2000)	**40.0**	78.3	39.8	2,697.0
Health expenditure, total (% of GDP) (1999/2000)	**5.1**	5.6	5.0	10.5
Hospital beds (per 1000 people) (1992)	**2.4**	2.7	2.4	9.3
Physicians (per 1000 people) (1994)	**1.7**	1.1	1.7	3.5
Immunization, measles (% of kid under 1 year) (2000)	**90.0**	70.9	85.4	75.0
Malnutrition prevalence, weight for age (% of kids under 5)	**10.0**	..	12.2	..
Mortality rate, under-5 (per 1000 live births)	**39.0**	84.7	45.3	6.0
Access to improved water source (% of population)	**75**	80	74	..
Life expectancy at birth (years)	**71**	69	69	77
Infant mortality rate (per 1000 live births)	**31**	33	36	4

Source: World Development Indicators Database and country at-a-glance (World Bank 2001)
Note: except the indicator illustrated, others are the data at the end of year 2001.

Health risk stems from the diseases and to a large extent depends on the pattern of disease prevalence in a country. The diseases can be classified into three types according their causes. The first type comprises infectious diseases, pregnancy diseases and diseases as a result of malnutrition; the second type incorporates cerebrovascular disease, cancer and mental diseases and other non-communicable diseases; the third refers to the intentional or accidental injuries. It is regarded that the shift of major diseases from the first type to second type results from transition of population reproduction type from at a high birth rate to at a low birth rate (Cai and Dou 2000). This need not be the case. To see why, the commonly encountered diseases and deadly diseases in China are tabulated in Table 3-9 and Table 3-10 respectively. It is obvious that infectious diseases (Acute pharyngitis coryza, influenza, and acute gastricism and entritis) remain among the top three common diseases that the surveyed people have contracted during the past two weeks before the survey time.

The deadly diseases, however, prevail in another way. As shown in Table 3-10, the top three deadly diseases in urban areas are malignant tumour, cerebrovascular disease and cardiovascular disease, in total these three kinds of diseases causes 65.7 percent of total deaths. For rural people, the most deadly diseases are respiratory disease belonging to the infectious diseases, followed by the three most fatal diseases for the urban in turn. The top three fatal diseases caused nearly 60 % of total rural deaths. Notwithstanding the shift of the pattern of diseases prevalence on the national level, the large majority of the rural population is still suffering the first type of diseases, which should have been prevented or treated effectively under the contemporary capacity of health care. The pattern of deadly disease prevalence in rural areas exhibits that the accessibility to health services remains a serious problem. On the other hand, the urban p attern s imilar to that of developed countries may indicate t remendous improvement of health care provision.

Table 3-9 Morbidity rate of five common diseases during the past two weeks in China in 2001

Code	Sum			Urban			Rural		
	Name of disease	MR	%	Name of diseases	MR	%	Name of disease	MR	%
1	Acute pharyngitis coryza	42.4	28	Acute pharyngitis coryza	44.6	24	Acute pharyngitis coryza	41.7	30.5
2	Influenza	14.4	9.6	Hypertension	15.5	8.3	Influenza	14.7	10.7
3	Acute and chronic gastricism	11.5	7.7	Influenza	13.5	7.2	Acute and chronic gastricism	11.7	8.5
4	Hypertension	6.6	4.4	Acute and chronic gastricism	11.2	6	Rheumatoid arthritis	5.2	3.8
5	Rheumatoid arthritis	5	3.3	Rheumatoid arthritis	7.2	3.9	Quinsy and tracheitis	4.2	3.1

Source: MOH 2002

Table 3-10 Mortality rate of ten leading deadly diseases in China in 2001

Order	Urban		Rural	
	Cause of death	%	Cause of death	%
1	Malignant Tumor	24.9	Respiratory Disease	22.5
2	Cerebrovascular Disease	20.4	Cerebrovascular Disease	19.0
3	Cardiovascular Disease	17.6	Malignant Tumor	17.7
4	Respiratory Disease	13.4	Heart Trouble	13.1
5	Trauma and Toxicosis	5.9	Trauma and Toxicosis	10.7
6	Internal System, Nutrition, Metabolite and Immunity Disease	3.2	Digestive Disease	4.1
7	Digestive Disease	3.2	Urinary Disease	1.5
8	Urinary Disease	1.6	Newborn Baby Disease	1.3
9	Mental Disease	0.99	Pulmonary Tuberculosis	1.2
10	Neuropathy	0.96	Internal System, Nutrition, Metabolite and Immunity Disease	1.1
	Total	92.0	Total	92.1

Source: NSB 2002

3.2.4 Retrospect of the rural health sector reform in recent two decades

The health sector reform implies sustained, purposeful and fundamental changes in the health sector (Berman 1995). The economic reforms beginning in the early 1980s greatly influenced China's health sector, revealing the weaknesses of the traditional health care system in adapting to the new socioeconomic environment. China's health sector reform is, thus, underway and has been a need to keep pace with the economic reforms. The health sector reform in past two decades mainly concentrated on two aspects: the health care financing system and the policy for health care providers. Owing to the tremendous differentiation between the urban and rural health sector, we then only focus on the reform of the rural health sector.

In rural areas, the reform in the agricultural production system led to the disintegration of the cooperative economic organizations that formed the basis of rural cooperative medical system. The government adopted a *laissez-faire* policy, and the rural population turned to paying for health care totally out of pocket. RCMS coverage shrank enormously; fewer than 10 percent of China's villages had the functioning RCMS (MOH 1999). Accompanying the breakdown of the health care financing system, the health care delivery system suffered too. A great many of doctors in village health stations had to leave for full-time farming or became private practitioners. Town health centers and county hospitals were also largely financed by fee-for-service, and out-of-pocket payment was actually in financial difficulty and their performance had deteriorated (Tang 1997). Ability to pay has become an important determinant of access to health care. Many households borrow money to pay medical bills and severe illnesses become an important cause of poverty.

In the period of 1980 to 1983 RCMS experienced a rapid collapse. Since then, the enrolment rate was leveling off at around 5 percent. In 1993 the Central Committee of CPC's "Resolution on a few issues of Construction of the Socialist Market Economic System" reaffirmed the major role of RCMS. Following the Resolution, the Center for Development Research of State Council, the Ministry of Health, the Ministry of Agriculture in collaboration with WHO started the pilot experimentation entitled "China's Rural Cooperative Medical System Reform" in 14 counties of 7 Provinces in 1994. The community-based health insurance schemes (CBHIS) had been set up in some rural regions.

In 1997 the Central Committee of CPC and State Council issued a more specific and authoritative document: "Resolution on Health Reform and Development". The resolution not only put emphasis on the role of RCMS in financing health care but also put forward some principal methods to develop RCMS. The Ministry of Health raised a series of measures in response to the resolution of State Council and also declared an ambitious agenda that by 2000 RCMS would be extended to 90 percent of the villages. However, the endeavors proved ineffectual as will be shown by a national survey conducted by the China academy of social science (CASS).

The *de facto* failure to achieve the target set by MOH in 1997 makes the problem in rural areas worse. In some regions the epidemic and endemic have been endangering the locals; the disease-induced poverty has been rising. This, together with the widening gap of income, may lead to social instability.

Moreover, the 16th National Congress of the CPC held in 2002 proposed an ambitious blueprint to establish the "overall well-off society" in China by 2010.Under the circumstances, the Central Committee of CPC and State Council approved and granted the "Resolution on further strengthening rural health work". As an integral part of the construction of "overall well-off society", the Resolution promises a new round of health reform in rural areas so as to increase the accessibility of the rural people to health care and hence health status to relatively high level by 2010. It points out again that rural health work remains a national priority. The objectives of this round of reform in rural sector are to (i) deepen the reform in rural health system, (ii) increase health inputs and optimize the allocation of health resources, and (iii) narrow the gap in health status and health resources between the urban and rural population.

The goal of the reform was to construct a well-equipped and -staffed health care delivery system and a new type of RCMS so that all peasants enjoy primary health care. This is a quite impressive target, but this plan is also a glaring admission of how bad the health care situation in rural China had become and is actually a compensation and correction of the failure of the rural health sector undertaking since the economic reforms. The Resolution elaborated on how to strengthen the rural health work from six aspects:

- Strengthening the public health work;
- Pushing the establishment of rural health care delivery system;
- Augmenting the health inputs;
- Building and perfecting rural cooperative medical scheme and health relief system;
- Strengthening the supervision of the market for the rural medicines and health services; and
- Reinforcing the leadership of rural health work.

The key of this reform, in short, is to augment health inputs for the rural population, especially in the middle and western regions. First of all, all levels of government are required to increase the health inputs at the growth rate of budgetary expenditure. Second, to establish rural cooperative medical scheme in all the rural villages by 2010; the central government would contribute 10 Yuan per year to health insurance fund for each rural participant in middle and western regions and local governments (provincial, prefecture, county and township governments) would contribute jointly at least the equal amount. This measure is expected to substantially prompt the revival of the former RCMS. The pilot RCMS are being implemented in each of middle-west provinces. Rural people are expected to be able to enjoy the welfare similar to their urban counterparts in the near future.

3.3 The Sichuan Province

3.3.1 Physical features, population, socio-economic development

Sichuan, in China's western hinterland, covers most of the Sichuan Basin. Surrounded by mountains, it enjoys a mild climate. Sichuan covers a vast area of 485 000 square kilometer, accounting for 5.1 percent of China's total area. Sichuan province is marked by its abundant natural endowment, and hence it has been known as "Land of Abundance" (*Tianfu zhiguo*). Sichuan is high in the west and low in the east in terms of topography. The Sichuan Basin, is one of the four largest in the country, covers an area of 165,000 square kilometer. However, the land-locked basin and the surrounding mountains make Sichuan relatively isolated from the outside world and hamper its social and economic exchange with other regions.

Sichuan was of a population of 84.37 million, of which the rural population accounted for 80.7 percent, according to the result of the fifth national census (NSB 2001). The population density was some 173.95 persons per square kilometer at the end of 2001.

In addition to its majority, the Han people, Sichuan is also inhabited by many ethnic groups, including 13 with at least 5,000 people each: the Yi, Tibetan, Qiang, Hui and so on. Sichuan has China's second-largest region inhabited by Tibetans as well as the largest region inhabited by the Yi ethnic group and the only region where the Qiang people live in homogeneous communities. The Yi ethnic group is of the largest population among all minorities in Sichuan.

The rural household of Sichuan had on average 3.93 persons and 2.69 labor forces. The dependency ratio was 1.46 (inclusive of the independents him- or herself). The family was characterized by the nuclear household. Gross annual income per rural person was 2946.51 Yuan and net annual income per rural person was 1986.99 Yuan, with 66.2 percent being cash income. Annual expenditure per rural person was 2494.07, Engel's coefficient[7] 0.55, according to the official statistics (SSB 2002).

Its economy is underdeveloped and agriculture remains a pillar industry. Sichuan is one of the major agricultural production bases in China. Grain, including rice and wheat, was the major products. Commercial cash crops include rapeseeds, citrus fruits, peaches, sugar canes and sweet potatoes. Sichuan also had the largest output of pork and the second largest output of silkworm cocoons among all provinces in China in 1999.

By 2001 the GDP has amounted to 442.2 billion Yuan, and GNP per capita 5279 Yuan, equivalent to 640 US Dollar[8], which was far lower than the national average, 890 US Dollar. The first, second and tertiary industry accounts for the GDP 22.2, 39.7 and 38.1 percent, respectively.

---

[7]By definition, Engel coefficient=food & beverage expenses/consumption expenditure× 100
[8]On the basis of the official exchange rate in 2001: 1 USD=8.26 Yuan

3.3.2 Health care system

Sichuan Province has 33,412 health institutions in total, not including health village stations. The major facilities comprise 854 hospitals at and above county level (inclusive of hospitals of Chinese medicine), 6,696 town or township health centers, 200 clinics, 188 maternal and child care centers and 219 sanitation and epidemic prevention stations. There are totally hospital beds 189,829, on average, 2.25 beds per thousand persons and totally 249,392 medical technical workers, on average, 2.96 professionals per thousand persons.

Since the mid-1990s, the fiscal spending on the health sector has been decreasing considerably in terms of ratio of health expenditure in total fiscal expenditure despite the stable increase in absolute amount. The health expenditure was 128 734 Yuan in 1995, taking up 7.1 percent of the fiscal spending while the health expenditure was 248216 Yuan in 2001, taking up 4.2 percent. In contrast, the rural household's health expenditure rose to 82.45 Yuan in 2001 from 29.35 Yuan in 1995. The share of rural household's health spending in consumption expenditures went up to 8.0 percent in 2001 from 4.9 percent in 1995. Plotting ratios of public and private health spending we can get a forficiform chart of tendency as shown in Figure 3-1.

The urban health care financing system in Sichuan is the same as those of other provinces. However, the rural areas are mainly dominated by out-of-pocket payment and the enrolment rate of RCMS is far lower than the national average. There are only a few counties where RCMS remains functioning.

Figure 3-1 Ratio of public & private health spending in Sichuan, 1995-2001

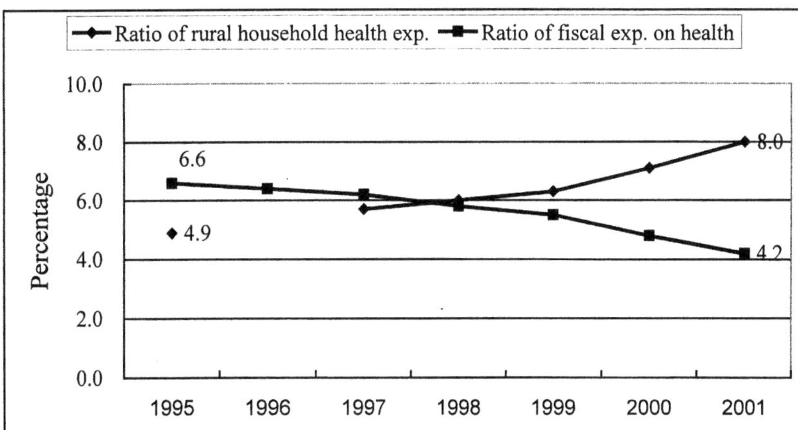

Source: computed from SSB 2002.

4. Demand for Health Insurance and Its Impact on Care-Seeking Behavior

As discussed in chapter 2, risk and uncertainty are inescapable in all walks of life. Illnesses and injuries befall people quite randomly, creating a health risk that in turn produces financial costs as people seek health services to recover health. Most people are risk averse and thereby there is a need for society to create institutional arrangements to reduce risk magnitude. Health insurance is one of these formal institutional arrangements. Obviously, demand for health insurance is primarily motivated by avoiding the fluctuation of financial costs due to illnesses and injuries. In other words, health insurance smoothes the fluctuation of health care costs and hence smoothes consumption. In this chapter, demand for health insurance is investigated and estimated, and its impact on health care seeking behavior is observed and explored next.

4.1 Theoretic framework—health, health care and health insurance

Health, health care and health insurance are so integrated that one cannot explore each of them effectively and meaningfully without taking account of their interactions. What people observe are demand for health care and health insurance, but what people really demand is health. Health care demand is simply a *derived* demand for an input to produce health. Costs of health care can be tremendous and uncertain to such an extent that normal income may not deal with it well without influencing normal consumption. Thus, health insurance is needed to reduce the fluctuation of costs of medical care. The researcher starts the exploration with the basic relationship between health and health care.

4.1.1 Health and health care

There exist two alternatives for modeling the way individuals make a choice of health care utilization and associated decisions. A simple approach is to treat health as one of several commodities over which individuals have well-defined preferences and to use orthodox consumer theory to investigate the determinants of demand (Phelps 1992; Jack 1999). A second approach is to analyze health care choices by employing an inter-temporal model of consumption decision while treating health as a stock variable within a human capital framework. Demand for health care is actually the investment in human capital (World Bank 1993; Folland *et al* 2001). Therefore in the most part of the thesis the orthodox static utility-maximizing framework will be adopted, but occasionally the idea of human capital investment will be applied as an alternative.

Health is the underlying demand of human beings and an important source of utility, while health care is one of the key productive factors of health. Grossman

(1972) prominently explored the idea of *health* as an economic good and showed how a rational economic agent would have a demand curve for health care services that *derived* from the underlying demand for health. Suppose that a bundle of other goods, *C*, and a stock of health (unobservable), *H*, then an individual's utility function may take the form

Eq. 4-1 $U = U(H, C)$

It is assumed that greater health and higher levels of other consumptions make the individual better off. That is, both are normal goods, then the marginal utilities of both *C* and *H* are positive. Using calculus, $UC = \partial U/\partial C > 0$ and $UH = \partial U/\partial H > 0$. The researcher also assumes that an increase in one coupled with a decrease in the other of a particular magnitude leaves an individual's welfare unchanged. Thus a standard indifference curve can be drawn representing preference between *H* and *C* with a slope of $-U_C/U_H$ (panel a of Figure 4-1). The total change in utility is given by $dU = dCU_C + dHU_H$ and $dC/dH = -U_H/U_C$ gives the usual economic notion of trade-off: *marginal rate of substitution* between C and H in producing utility.

The consumption of other goods does not only bring utility but also produces health. Thus, health care and other goods are productive factors of health if health is also looked on as a special production.

Eq. 4-2 $H = f(m, C)$

In this equation, the respective contribution to health of health care services and other consumption can be separated. If health status rises slowly and consumption of goods increases relatively quickly, then it illustrates the decline in use of health care services. The marginal product of *m* is the partial derivative $\partial H/\partial m$. Since *m* can be converted into *H*, that relationship can be embed in the utility function, so that

Eq. 4-3 $U = U(H, C) = U[f(m, C), C]$

Using the chain rule, it can be easily found the relationship between health care and utility

$$\partial U/\partial m = (\partial U/\partial H)(\partial H/\partial m) = UH\,gm$$

Subject to income *I*,

Eq. 4-4 $I \geq P_C C + P_m m$

Thus, the individual tries to maximize the utility under the constraint of income, given the preference of consumption. When the individual falls ill, costs of maintaining the original status increase and the boundary of the budget set swings inward (panel b of Figure 4-1). The effect of falling ill is just the same as having the price of health increase since the patient needs more health care.

A close look at the production function is given. It is presumed that more *m* produces more *H*, that is, the marginal product of medical care is positive, $f'(.)>0$. Also, in common with other economic phenomena, it is presumed that the marginal product is diminishing with more *m* input, $f''(.)<0$. After stipulating these characteristics, the production function can be illustrated with a concaved curve as shown in panel a of Figure 4-2 that is quoted from Folland *et al* (2001). Note that the production curve is a bit different from the orthodox production curve that is seen in most literatures of microeconomics. The health production curve is always concave and has no turning point from the increasing marginal product to the decreasing marginal product. Namely, the marginal product curve is always downward sloping[9], as shown in panel b of Figure 4-2. Empirically, due to the limitation of data, the production function is built on a highly aggregate level. For example, one can construct the production function of health (measured by life expectancies of countries or provinces) on gross medical expenditures per capita.

Figure 4-1 Indifference curve of health and consumption

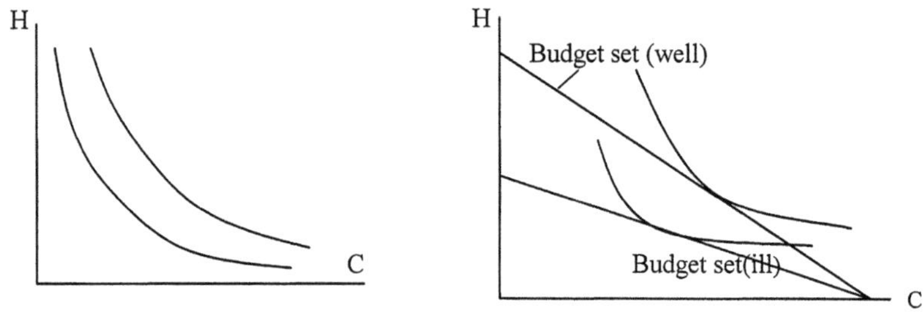

(a) (b)

Source: Phelps 1997

Figure 4-2 The production curve and marginal product curve

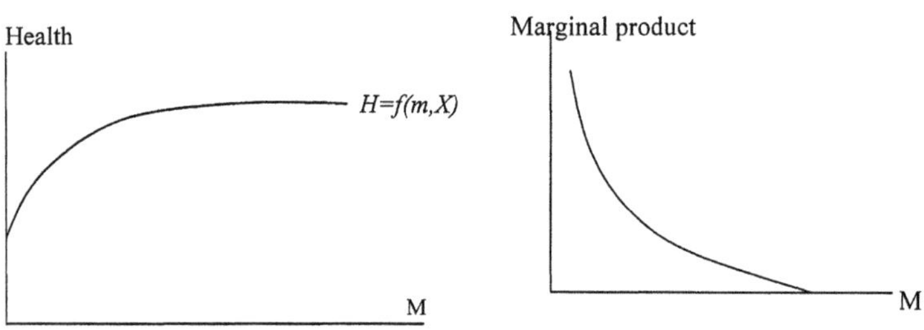

(a) Production curve of health (b) Marginal product curve of health care

Source: the author

Health production with health care occurs discretely. One demands health care only when he or she falls ill (this is not always the case, e.g. one may need

[9] Because it seldom makes sense to operate in a realm in which returns are increasing, we think about productions with decreasing returns.

regular check-ups even if one does not feel uncomfortable). An illness can be economically interpreted that the stock of health is lower than certain threshold, H^*, and the depreciation rate δ_0 for the moment is far higher than the average δ_t of a year. The function of health care is to raise the rate of depreciation δ_0 to δ_t and restore the stock of health to above H^*. Thus, health care is an input needed when the stock of health drops sharply and the rate of depreciation goes up suddenly due to illness.

On the basis of the health production function, the main determinants of health can be identified. Health care is the key productive factor of health and no commodities have the perfect substitution for it. Thus, the quantity of health care used is the most important determinant of health. Although the medical care is taken as homogeneous for the convenience of analysis, the quality of medical care is variable and takes a lot of forms. Outpatient and inpatient services are two common categories of health care; services provided by a surgeon are completely different from those provided by nurses.

Consumption goods certainly influence the production of health, either in the form of health input directly or by affecting the efficiency of production indirectly. For instance, food, clean and safe water and air, cloth, and physical building are non-medical inputs. Food, and water directly enter into the digestive canal and are converted into nutrient substances necessary to maintain the health status when no illness occurs. When one is ill, various nutrients derived from food are even more important. On the contrary, insecure food, contaminated water and air may enhance the incidence of some diseases and sharply increase demand for health care to compensate for the loss of health due to the increased rate of depreciation.

Physical environment (e.g., homes, workplaces, as well as streets and highways) is a determinant of health by affecting the efficiency of health production. The common feature of these factors is that an individual is less able to adjust the level of investment and has to depend on households, employers, communities or even governments. Therefore, physical environment can be considered given for an individual in the short run, though its impact on health outcome varies from person to person. Workplaces are places where the employers pursue profits while the employees earn their living; it is estimated that people spend about one fourth to one third of their lifetime here. A safe and comfortable workplace not only reduces the probability of injury caused by working accidents and thereby decreases demand for health care, but also lowers the probability of falling ill. With a good infrastructure (transportation and communication means) one has easier access to health care; it takes less time for him or her to purchase health care.

Demographic attributes like age, gender and education also determine health outcome of an individual by affecting health productivity. As is observed, health deteriorates as one ages. This process can be thought of as a decline in health productivity. Compared with a youngster, an old man has to use much more medical care to produce the same level of health. Where gender is concerned, it is recognized that females are more health productive than the males because they can maintain health above a minimum level and lead a longer life. Under the Grossman approach, better-educated people tend to be economically more efficient producers of health because they understand more of the technology of health production or have more of the required knowledge to stay healthy.

However, consumption of some other goods and services may have potentially negative effects on health. These effects involve exposure to pollution, increased likelihood of motor vehicle accidents, over-consumption of alcohol, smoking, indulgence in unsafe sex, and consumption of unhealthy foods. To make a distinction of health improving consumption from health damaging consumption, the production function can be written

Eq. 4-5 $$H = f(\overset{-}{X_B}, \overset{+}{X_g}, \overset{+}{m})$$

where X_B represents beneficial goods and X_g bad goods. We should, however, be cautious to argue that people should not consume bad goods because of their effects on health. People choose to consume bad items even when they are totally aware of the side effects on health. This might be understood by assuming that those people are utility maximizer rather than health maximizer, or/and they are less risk averse than others. The role of bad goods in militating against health tends to be underestimated; these behavioral choices dominate a person's health far more than the medical system (Phelps 1997).

Though income makes an impact on health through the financial constraints of health production, income is empirically a sound direct explanatory factor of health. It is expected that consumption of health-improving goods and services rise accordingly as income rises. Thus, the higher income one has, the healthier one may be, other conditions held constant. Income can be divided into two parts assuming that one has no initial wealth: one is the earning income that is quantitatively equal to work time multiplied by wage rate; the other the net transfer. Wage rate impacts health by adjusting one's time trade-off between labor and leisure[10]. An individual who has a higher wage rate tends to have a high optimal level of health. Transfer, especially transfer aimed at health care purchase, will contribute to the health improvement of an individual.

The above efforts to identify the determinants of health have been made simply because it is believed that determinants of health are potentially the determinants of demand for health care and health insurance. This basic identification will facilitate the following identification of determinants of demand for health insurance.

4.1.2 Interaction of health insurance demand and health care demand

Health care and health insurance markets are interrelated so closely that we cannot meaningfully explore health care markets or health insurance markets

[10] A basic assumption is that leisure time is used to produce health and home good for health.

separately. First, a look at the effect of health insurance on demand for health care is given. On the one hand, health insurance lowers effective price of health care. On the other hand, it increases a welfare loss. If health care is a normal good, providing people with a health insurance policy should increase their utilization of health care. The additional health care consumed as a consequence of lower patient-borne costs is of lower value than goods that otherwise would have been consumed under full price conditions. The consumer faces a consequent welfare loss (Phelps 1997). The welfare loss is none other than the decreased marginal value created by the insurance-induced demand for health care. The substitution effect of health insurance may also create a price distortion or inefficiency in the market for health care. The policies with different features (coinsurance, deductibles, upper limits on coverage) have different mechanisms in elevating demand for health care.

Economic theory of demand for health insurance relies on the assumption that demand for health care is price elastic (Feldstein 1973). The greater the price elasticity, the more insurance can be expected to induce purchases of low-value care. However, the coinsurance mechanism can mitigate this over-utilization, more precisely coinsurance mechanism alters the consumer's choices for health care in this way. Suppose that demand for health care is taken a linear approximation, without insurance the demand curve of health care can be written

$$M_{ni} = a_0 + a_1 I + a_2 P_m + a_3 L$$

where *I, P, L* denote income, price and level of illness, respectively. As shown in Figure 4-3, D_2 is the demand curve without insurance policy. The price that consumers pay for medical care with a simple insurance policy is CP_m, where C is the coinsurance rate and Pm is the demand price. The quantity of health care demand increases to m_4 from m_3 with no insurance policy. Downward along the out-of-pocket demand curve D_2, the intersect point with the lowered price CP_m line is just the new quantity demanded. It is more useful to identify the demand curve with respect to the market price. The straight line D_2' that links *N* and *E* is the demand curve with respect to the market price[11]. The insurance costs *R* per year, thus the budget constraint for the consumer with insurance is

$$I - R = P_X X + CP_m m$$

Accordingly, the demand curve with insurance (M_{wi}) is

$$M_{wi} = a_0 + a_1(I - R) + a_2(CP_m) + a_3 L$$

The price elasticity of the demand curve is reduced from $a_2 Pm/M_{ni}$ to $a_2(CPm)/M_{wi}$, where $1/M_{ni} > C/M_{wi}$. Thus, due to the basic features of co-payment, the health insurance will make demand for health care less elastic.

Apart from the coinsurance rate, there are still other features such as

[11] The detailed process of obtaining the demand curve with respect to market price can be seen in Folland *et al*, 2001, p173.

deductible and maximum payments limit. The effect of a deductible on demand has two possible impacts. A relatively small deductible will have little effect on the individual's utilization. A large deductible makes it more likely that an individual is self-insured and consume the amount of care he or she would have purchased with no insurance. The presence of deductibles discourages frivolous claims or visits and also makes insured people more aware of the results of their action. The effects of a maximum payment limit on insurance are the reverse of a deductible—it leaves really serious medical events *uninsured.* More popularly, many private insurance policies now have a *stop-loss* feature that places an upper limit on out-of-pocket spending by the consumer, even if there are co-payments for some care.

On the contrary, demand for health care affects demand for health insurance. The influence of demand for health care on demand for health insurance is realized through its elasticity. The expected welfare loss depends directly on the elasticity of demand for medical care. The correlation between health care demand and health insurance demand is that the more price responsive (price elastic) demand for health care is, the less desirable it is to insure against the risk with normal types of health insurance policies. The reason for this is that large price responsiveness creates bigger welfare losses in demand for health care for any given level of insurance coverage (Phelps 1997). Suppose that demand for health care is taken a linear approximation, as shown in Figure 4-3. The demand curve D_2 is less price responsive, so if one purchases an insurance policy with co-payment rate *C*, the increase in demand on demand curve D_2 is only to quantity m_4 from m_3 without an insurance policy. The welfare loss is the small shaded triangle *B*. However, if the demand curve is much more elastic (price responsive), as shown by D_1, then the quantity demanded with an insurance policy rises to m_2 from m_1 and the welfare loss is the shaded triangle *A*, obviously much larger than *B*. An intuitive illustration of this idea is that people tend to buy hospital and surgical services insurance that are price inelastic, while not purchasing dental services coverage that is quite price elastic.

Figure 4-3 Impact of health care demand elasticity on demand for health insurance

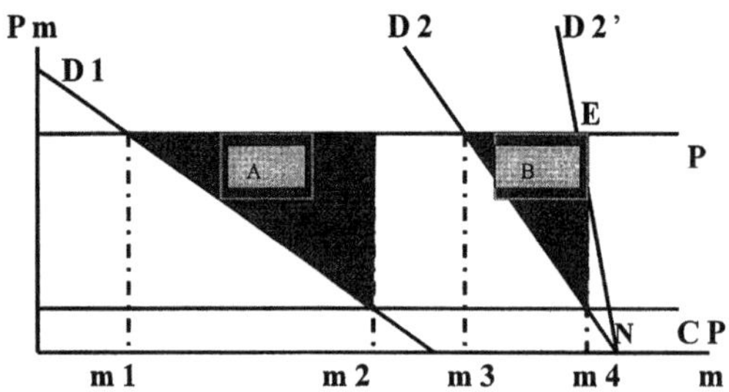

Source: Phelps 1997

4.2 Demand for health insurance

4.2.1 Modeling demand for health insurance

The nature of rural health insurance

As stated in Chapter 3, there are basically two types of health insurances in rural areas; namely, private health insurance (PHI) and rural cooperative medical scheme (RCMS). Private health insurance policies in rural China can be purchased to cover a specific member. RCMS are usually participated in on a household basis and the participation of a portion of selected household members is institutionally forbidden. These two types of health insurance forms differ substantially in supply. The suppliers of PHI policies are large insurers with a huge number of branches nationwide, whether in urban or rural areas. Consequently, risk pool is quite large and the insurance policy is open to anybody. In contrast, RCMS is supplied by local community and risk pool is quite small. The villagers residing in a community without supply of RCMS are surely not members of RCMS. Where the correlation between them is concerned, the researcher argues that the membership of RCMS would influence the purchase of policies, but the reverse is not true, because RCMS has a low premium and its coverage is broad compared with PHI.

Although there are a number of private health insurance suppliers in China's health insurance market, there are basically only three insurers in rural market. China Life Insurance Company dominates the rural health insurance market. Each company offers a number of health insurance products, but only two are presently most common for the rural inhabitants. One is for adults named 'major diseases benefits' and the other for pupils and kindergarten kids called 'pupil health and accident insurance'. The "major disease benefits" has an element of saving; a certain amount of money could be repaid in case the insurants have not infected the diseases within the insurance term stipulated in the insurance contract. The benefits of an insurance policy lie mainly in the compensation for the likely medical costs of certain types of inpatient services in county hospital. The benefits of RCMS are similar to PHI, but members of RCMS can seek the health services in all public health facilities so long as they follow the referral procedure. To deal with the issue of moral hazard, coinsurance rate is specified from 5 to 20 percent for PHI and 40-70 percent for RCMS. Deductibles and maximum payment limits for a specific disease are also employed to contain the costs of care. In this market the choice of insurance is essentially a discrete choice between some and no insurance. The choice of coverage is secondary.

A static model of demand for health insurance

There are two possible states of the world ($x \in \{w, s\}$): state (w), 'well' having the probability ($1\text{-}p$) and state (s), 'sick' having the probability p, $0<p<1$. Consumers differ only by their probability of sickness. Each consumer's loss due to sickness is $L(s)= \eta M$, M is medical care costs and η is the probability of seeking care, both are assumed unaffected by the presence of health insurance. Obviously, $\eta \in [0,1]$, 0 represents not seeking care and 1 seeking care. $L(w)=0$ in state s.

Having public information and without having transaction costs, a risk neutral private monopoly would offer health insurance policy characterized as (α, β), with coverage (net of premium) (β) for an insurance premium (α) such that a consumer would be indifferent between purchasing the policy and having no insurance (Stiglitz 1977). In other words, the private monopolist maximizes his total profit over α, β. If W_1 denotes the wealth in 'well' state and W_2 the wealth in 'sick' state, then the expected utility of the individual is as follows.

Eq. 4-6 $$EU(p,\alpha,\beta) = (1-p)U(W_1) + pU(W_2)$$

where $U(.)$ is von Neumann and Morgenstern expected utility function with $U'\geq 0$, $U''(.)\leq 0$, namely the agent is risk averse and degree of risk aversion is R. The expected utility of policy purchase is:

Eq. 4-7 $$EU(p,\alpha,\beta|M,\eta,R) = (1-p)U(W-\alpha) + pU(W-\eta M+\beta)$$

The expected utility of not purchasing policy is:

Eq. 4-8 $$EU(p,0,0|M,\eta,R) = (1-p)U(W) + pU(W-\eta M)$$

For a given p, α, β, η, W, M, and R. Insurance will only be chosen if and only if:

$$EU(p,\alpha,\beta|M,\eta,R) \geq EU(p,0,0|M,\eta,R)$$

In the absence of asymmetric information and under the standard assumptions of insurance models, a Pareto optimal solution is characterized by full insurance coverage for all individuals in each class of risk. Each insured sets his optimal consumption level according to his/her certain wealth. No other financial institution is required to obtain this level of welfare. The two standard theorems of welfare economics hold and market prices of insurances are equal to the corresponding social opportunity costs.

However, in the presence of asymmetric information, adverse selection will emerge. If all insurers have imperfect information on individuals' risks, health insurance market may not exist, or if it exists, it may not be efficient (Akerlof 1970). Even in monopoly health insurance market, equilibrium may not exist.

Within the theoretic framework, a rational economic agent may take into account the following relevant factors before (s)he makes the decision to buy an insurance policy or participate in a health insurance scheme: probability of illness (p), degree of risk aversion (R), magnitude of medical care costs (M), probability of seeking care (η), premium of insurance (α), insurance coverage (β), income (W). Apart from those, individual and community characteristics also influence demand for health insurance: Age, gender, education, occupation, employment sector, level of industrial and commercial development, and province are all associated with being insured.

Empirical specification of the estimation model

Based on the theory discussed above, demand for health insurance and demand for health care interact with each other. On the one hand, being insured by a policy or being member of RCMS will increase demand for health care; on the other hand, the expected large costs of health care will increase demand for health insurance. However, rational agents can only use the past data of costs of health care to predict the future demand for health care and then come to a decision whether to buy insurance or not, that is, the need for health care cannot affect the contemporary demand for health insurance. Therefore, in the case of cross-sectional data, the impact takes place unilaterally: Demand for health care, *M*, is a function of a set of exogenous variables, *X*, and participation in health insurance, *Y*. *Y*, however, can only be a function of a set of exogenous variables under the circumstances because historical observations are unavailable in the cross-sectional data.

For simplicity, demand for health insurance (scheme), Y^*_j, is taken a linear approximation again, which is function of some of the exogenous variables *X* as follows:

Eq. 4-9 $Y^*_{ji} = \gamma X_{ji} + u_{ji}$

where Y_{ji} is the dependent variable that shows the occurrence of purchasing a health insurance policy by an individual *i* in household *j*. X_{ji} is a vector of household and individual variables that impacts the decision to purchase health insurance.

In practice, Y^*_{ji} is unobservable, what we observe is a dummy variable defined by

Eq. 4-10 $Y=1$ *if* $Y^*_{ji}>0$ *(if individual i in household j is insured privately)*
$Y=0$ *otherwise*

From the relations between Eq. 4-9 and Eq. 4-10, according to Maddla (1983), we get

Eq. 4-11 $\mathrm{Prob}(Y_{ji}=1) = \mathrm{Prob}(u_{ji} > -\gamma X_{ji}) = 1 - F(-\gamma X_{ji})$

where F is the cumulative distribution function for *u*.

In this case the observed values of *Y* are just realizations of a binomial process with probabilities given Eq. 4-11 and from trial to trial (depending on X_{ji}). Hence, the likelihood function is

Eq. 4-12 $L = \prod_{yji=0} F(-\gamma X_{ji}) \prod_{yji=1} [1 - F(-\gamma X_{ji})]$

The function form for *F(.)* in Eq. 4-12 will depend on the assumptions made about u_{ji} in Eq. 4-11. If the cumulative distribution of u_{ji} is logistic, we have a *logit model*. In this case,

Eq. 4-13 $F(-\gamma X_{ji}) = \dfrac{\exp(-\gamma X_{ji})}{1+\exp(-\gamma X_{ji})} = \dfrac{1}{1+\exp(\gamma X_{ji})}$

$\therefore$ Eq. 4-14 $1-F(-\gamma X_{ji}) = \dfrac{\exp(\gamma X_{ji})}{1+\exp(\gamma X_{ji})}$

So there is a closed-form expression for F in this case, because it does not involve integrals explicitly. Not all distributions permit such a closed-form expression. In a probit model it is assumed that u_{ji} are IN(0, σ^2). In this case,

Eq. 4-15 $F(-\gamma X_{ji}) = \int_{\infty}^{-\gamma Xji/\sigma} \frac{1}{\sqrt{2\pi}} \exp(-\frac{t^2}{2})dt$

It can be seen from equation Eq. 4-15 and the likelihood function Eq. 4-12 that we can only estimate β/σ, and not β and σ separately. Hence, the researcher might assume $\sigma=1$.

The cumulative normal distribution and the logistic distribution are very close to each other, except at the tails. Thus, it is unlikely for us to get very different results through using equation Eq. 4-13 and Eq. 4-15. That is, the logit or the probit method, unless the samples are quite large so that enough observation are at the tails. However, the estimates from the two methods are not directly comparable. Ameiya (1981) suggested that the logit estimates be multiplied by 1/1.6=0.625, saying that this transformation produces a close approximation between the logistic distribution and the standard normal distribution.

4.2.2 Datasets

The study is based on a detailed Health Insurance Survey in Rural Sichuan (HISRS) conducted by the researcher. A National Survey on Rural Health System (NSRHS) conducted by Institute of Economy, Chinese Academy of Social Science (CASS) at the beginning of 2001 is used as a contextual or complementary data when Sichuan province's data is not sufficient for some parts of the analyses. The data of the Health Insurance Survey in Rural Sichuan is expected not only to be sufficient for the estimation of demand for health insurance in rural Sichuan, but also to allow for some close analyses on such issues as health risk coping strategies of rural households. The role of extended family as well as the performance of the existing RCMS will be addressed in the forthcoming chapters. Sichuan province is one of six sample provinces within the NSRHS.

4.2.2.1 Sampling method of the Health Insurance Survey in Rural Sichuan

The HISRS is aimed at collecting data, which is needed for the solutions to the problem raised in chapter one. Both the household-level and organization-level data were collected. At the level of household, the survey centered on data of demographic characteristics, health status, health care seeking behaviors, income and households' social relations that share risks of illness with the respondents' households. In addition, to explore the risk coping with the risk of illness within the extended family the situation of reciprocal transfer such as labor, gifts and cash are investigated. At the level of the insurance providers, the data on premiums, benefits packages and other characteristics of health insurance products were collected; many aspects of the operation of RCMS by local communities, for example, the characteristics of community, the sources of funds and the magnitude of transaction costs were collected.

A multi-stage sampling procedure was used to select townships and villages for the study. First, five counties representing different level of development, geographical location and ethnic groups were selected. Second, two towns (or townships) in each of the counties were chosen: (a) one middle-income town was randomly selected, (b) another was chosen totally randomly. Third, one village in each of the townships was selected randomly. In total, three hundred households fell into the sample. The researcher also investigated the insurers that were providing health insurance for rural people in Sichuan province. The sampling stage and size is shown in Figure 4-4.

4.2.2.2 Process of survey

The first step of the survey was to translate the questionnaire from English into Chinese. Enumerators were then recruited from senior students specializing in agricultural economics or industrial-commercial management at Faculty of Economy and Management, Sichuan Agricultural University. Apart from their excellent comprehensive performance in study and character, another criteria was that they should come from the sampled county so that each can speak the local dialect, which makes it easy to approach the respondents and communicate with them. A very important link of the chain is for interviewer to receive training of investigation. Training contents include research problems, objectives, research questions, significance of the research as well as how to implement the inquiries. Together with interviewers, the researcher carried out a pilot survey in six households, each interviewer went through an interview with one household independently and then they were required to summarize the lessons in the survey and exchanged their useful communicative skills in a meeting.

Enumerators, however, needed to select the sampled households by themselves after they got the name lists of household heads from the director of village commission. Counties, towns and villages were selected in advance of their departure for field survey. Enumerators visited the sampled households twice: first they wrote down the basic information of households and made an appointment for the second relatively long interview. On the second visit, they went through the whole survey.

Figure 4-4 Method of sampling households

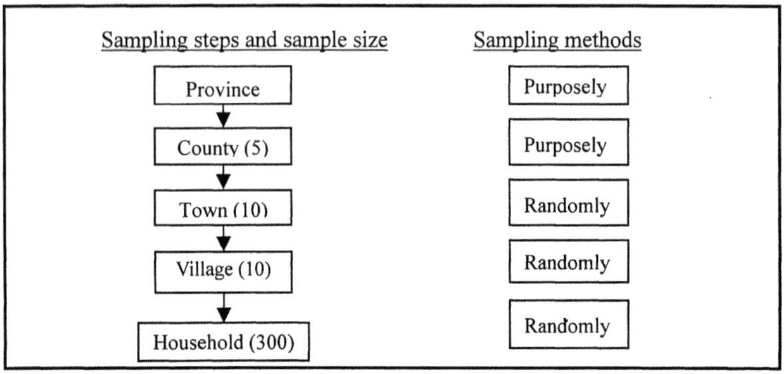

4.2.2.3 China dataset

A fraction of data extracted from the National Survey on Rural Health System (NSRHS) contains 4 questionnaires at the county, township, village, and household level, respectively. There are 6 sample provinces all over the country and in each of the provinces there are 6 sample counties, then only one town (or township) was selected in each of the counties and one village in each of the towns (or townships) was selected randomly. The geographic location of the provinces is shown in Figure 4-5. A brief description of the six sample provinces is tabulated in Table 4-1. Sichuan province is the key sample province.

Figure 4-5 Location of the sampled provinces in China dataset

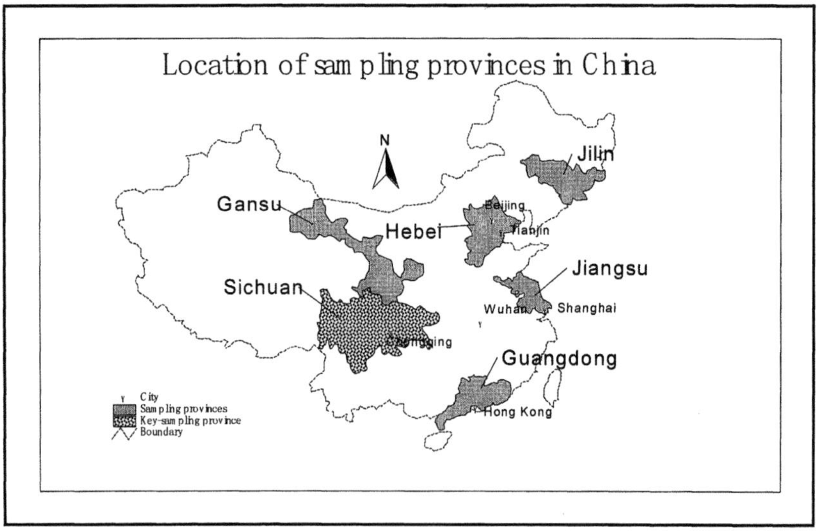

Table 4-1 Sample provinces and sample size of the NSRHS data in China

Provinces	Location & development level	Number of counties	Number of households
Jiangsu	East and developed	6	489
Guangdong	Southeast and developed	6	520
Jilin	Northeast and less developed	6	430
Hebei	Middle and less developed	6	556
Sichuan	Southwest and underdeveloped	6	300
Gansu	Northwest and underdeveloped	6	210
Total	-	36	2505

Source: NSRHS by CASS

4.2.2.4 Description of sample counties in Sichuan, China

As mentioned above, the sampling criteria in HISRS are geographic location, development level and ethnic group. In terms of geographic location, this criterion attempts to cover a geographic variety that represents plains, hills and mountainous areas. The development level concerned mainly refers to rural average income per capita, the intention is to select counties that represent different levels of economic development. Sichuan is a province of many ethnic groups. Over 5 percent of the total population can be accounted for by minorities. So the sample should contain a minority county. Taking into account these criteria mentioned, Jiange, Fushun, Yanyuan, Hanyuan and Dongpo were selected Figure 4-6. The detailed indication of geographic location, economic development, and ethnic group of the 5 counties are presented in Table 4-2.

Figure 4-6 Sample counties of Sichuan Province

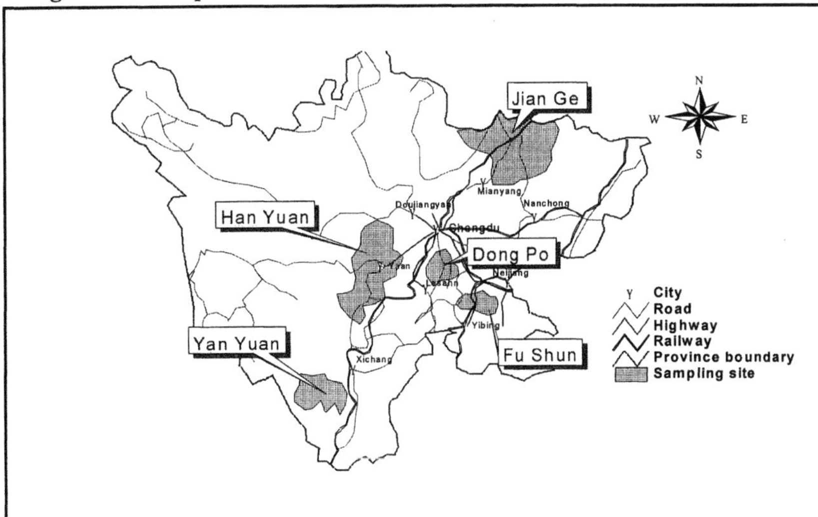

Table 4-2 Areas, population, and average income per capita by county, 2001

County	Area of Land (Km^2)	Pop. (0 000)	Pop. Density (Ca./	GDP per capita (Yuan)	Average income per farmer (Yuan)	Main nationality	Geographic location
Jiange	3204	67.01	209.14	1994	1397	Han	Hill
Fushun	1603	121.0	754.83	3186	2728	Han	Low Hill
Yanyuan	8398	31.9	37.99	1955	1132	Yi	Plateau
Hanyuan	2349	34.7	147.72	3212	1695	Han	Mountain
Dongpo	1331	81.5	612.32	5290	2524	Han	Plain
Province	485000	8436.6	173.95	5250	1903.6	Han	

Source: SSB 2002

Considering the research objectives the researcher is more interested in such health indicators as the average number of health facilities and health status at the county level (Table 4-3). Dongpo, the plain area and most developed county, has the best health infrastructure, whether in terms of hospital beds per 1000 persons (1.83) or in terms of medical technical professionals per 1000 persons (2.61). The least medical technical professional per 1000 persons (1.54) is in Yanyuan county, the minority and plateau area. However, Fushun county, which is most populous and in a low hilly area, has the lowest hospital beds per 1000 persons (0.86).

Table 4-3 Health status and health infrastructure by county, 2001

Name of counties	Jiange	Fushun	Yanyuan	Hanyuan	Dongpo
Infant mortality rate/1000 live births	39.4	16.4	-	-	16.0
Life Expectancy	69.0	69.7	-	-	71.1
Number of towns or townships	57	45	33	40	33
Number of hospitals	62	79	52	45	72
Number of hospital beds	996	1041	416	497	1495
Hospital beds/1000 persons	1.49	0.86	1.30	1.43	1.83
Medical technical professionals (MTP)	1298	2135	492	588	2129
MTP per 1000 persons	1.92	1.76	1.54	1.69	2.61
Number of consulting (10,000 person-time)	95.12	35.33	9.00	23.34	214.00

Source: NSRHS by CASS and SSB 2002

4.2.3 Description of factors affecting demand for health insurance

Private health insurance (PHI), rural cooperative medical scheme (RCMS) and the prepayment schemes were investigated in Health Insurance Survey in Rural Sichuan (HISRS). Results indicate that 144 persons are policyholders of private health insurance, accounting for 11.8 percent of all the sample individuals; 200 persons (or 16.4%) were members of RCMS and 33 persons (or 2.7%) members of the prepayment schemes. The prepayment schemes are not further analyzed in this study since they are far different from health insurance in terms of definition. A variety of factors affecting the two types of health insurances are identified in turn in this section.

4.2.3.1 Factors affecting the purchase of private health insurance policies

An attempt is made to identify factors that affect the purchase of *private health insurance* policies. The possible factors can be classified into 4 types: individual characteristics, household attributes, health risk related factors, and regional differences. The method of *contingent tables* is used in this section to present the number and percentage of policyholders as well as the results of the bivariate significance test.

A. Individual characteristics

For the research six indices were designed to reflect demographic characteristics, which may exert an influence on the decision of policy purchase. The six indices, as indicated in Table 4-4, are gender, age, education level, marital status, occupation and social position (whether or not an adult is a village cadre), respectively. Bivariate significance test shows that four of the individual attributes, age, education, marital status, and occupation, exert a significant influence on policy purchase.

Individuals at the *age* of 0 to 14 are most likely to be insured by health insurance policies. With the increase in age, individuals become less likely to be insured, except that the group at age of 30 to 44 has the least likelihood of policy purchase ($p<0.01$). There may be three facts underlying this tendency: first, the age group of 30 to 44 usually has a quite low probability of being ill; second, the parents traditionally place much importance on children; third, the insurance premium level for the younger is low relative to that for the adults, hence it is affordable for many families.

Education levels, if roughly categorized, can be no education, primary school, junior high school, secondary high school and college and above. This categorization describes how much time a person has invested in education (input aspect) and is different from the classification into illiterate and literate (output aspect). Results demonstrate that the more educated a person is, the more likely he/she is to be insured. Only 2.3 percent of those who have never gone to school are insured, the purchase rate of insurance policies among people who have received some primary education (1 to 5 years) increases largely by 9 percent. The purchase rate does not vary much from primary education to junior high school, while secondary high school graduates have a marked increase in purchase rate; needless to say, people who graduated at college are much more likely to purchase a health insurance policy (35.4%)than the previous education groups. All the differences are highly and statistically significant ($p<0.01$).

Marital status also affects the purchase rate of PHI policies significantly. Non-adults have the highest purchase rate, 20.2%; among the adults the single group has the highest purchase rate 15.9 percent, which is more than 2 times as much as the married group. However, the divorced and widowed groups are less likely to purchase policies compared with the single group ($p<0.01$), whose purchase rate is roughly consistent with that of the group at the age of 15-29.

Occupation is the last individual characteristic that significantly affects purchase of PHI policies. The purchase rates of groups of preschool kids and pupils, workers in village and town enterprises (VTE) and village teachers are all over 30 percent, whereas farmers, migrant workers and servants have far lower purchase rates. A statistical test verifies the significance of the difference among all the groups.

The factors of gender and social position do not exercise a significant influence on the decision of policy purchase. Statistics demonstrate that males and females have the same probability of being insured, and so do the local village cadres and common villagers.

Table 4-4 Cross table of individual characteristics vs. PHI policy purchase

Variable		Count	Purchase rate %	Significance	Expected sign
Age	0-14	52	19.3		
	15-29	46	13.9		-
	30-44	18	5.7	0.001	
	45-59	21	9.8		
	60-most	5	7.6		
Education	No education	8	2.3		
	Primary school	33	11.5	0.001	+
	Junior school	30	12.0		
	Secondary school	40	16.8		
	College and above	29	35.4		
Marital status	Single	20	15.9	0.001	
	Married	53	7.4		
	Divorced	1	9.1		
	Others (Non-adult)	70	20.2		
Occupation	Farmer	25	3.9	0.001	
	Worker in VTE	9	36.0		
	Village teacher	5	38.5		
	Migrant worker	4	4.3		
	Service sector	2	5.7		
	Others (pre-school kid and student)	46	31.5		
Relation to head	Head	30	12.5	0.001	
	Spouse	16	7.1		
	Kid	87	23.1		
	Others	11	9.2		
Gender	Male (=1)	77	12.3	0.574	+
	Female	67	11.3		
Social position	Village cadre (=1)	0	0	0.177	+
	Common villager	72	8.8		

Source: ZEF survey 2002

B. Household characteristics

Income, value of harvest and poverty status

For the study income is measured in three ways: annual income of household in cash, annual harvest value and perception of wealth status of the household in comparison to other villagers. The first two are integral components of the total household income. Cash income consists of off farm incomes, profits of sideline business, wages of part time jobs and other cash incomes (e.g., sales of livestock). Harvest value is computed by multiplying the quantities of planting agricultural products (mainly accounted for by food grain) by their respective market prices. High harvest value does not mean high cash income. Table 4-5 shows that health insurance policyholders have significantly higher annual cash income than do the non-policyholders, whereas the policyholders have on average lower annual harvest values. In statistic terms there is a positive relation between cash income per capita and policy purchase, and a negative relation between the harvest value and policy purchase. Statistical tests for interval variable (income or harvest value) and nominal variable (insurance status) indicate that both harvest value and income of households are intimately related to the purchase of PHI policies. The influence in opposite directions between income in kind and income in cash on policy purchase might be explained by the fact that the local rural households are to a large extent self-sufficient and the food grain market is underdeveloped, thereby it is kind of difficult to translate the harvest into cash income.

In addition to income in cash and in kind, the respondents were also asked whether their own households are poorer, average or richer in comparison to the others in the local village. More than 48 percent of respondents reported being 'on average'. The rate of policy purchase in the group of 'on average' is the highest, but this is not statistically significant. Unexpectedly the result on the basis of subjective measurement of wealth is not consistent with the result based on real income. Compared with the subjective measurement of income, the real income should be more reliable.

Table 4-5 measurement of income of policyholders and non-policyholders

	Policyholder	Non-policyholder	Overall	Significance
Annual harvest value (Yuan per household)	1449.28	2420.57	2305.64	0.01
Annual cash income (Yuan per household)	20725.76	9292.68	10785.30	0.01

Source: ZEF survey 2002. Note: Unit: Yuan≈0.125Euro at the time of survey

Physical barrier

In the questionnaire, two indices have been designed to measure the geographic barrier to the health insurance purchase: home's distance from the town center and walking time to the highways leading to the town center. In Table 4-6 it can be seen that non-policyholders live on average 1.6 kilometer further from the town center than do policyholders; meanwhile, non-policyholders have to spend almost 2 minutes more to walk to the highways leading to the town centers. The difference of 'walking time' between policyholder's and non-policyholders takes up 16 percent of non-policyholders' walking time, while the difference of home distance takes up 57% of the non-policyholders' home distance. We believe that the 'distance' captures more information on physical barriers than the indicator of 'walking time to highway' because the rural highways are usually of bad quality and the vehicles are quite few.

To generally examine the relationship between private health insurance purchases and the physical barriers, the sample households are divided into six groups in terms of home's distance from the town center and walking time to highway. Both the number of policyholders and density of private health insurances by various groups are shown in Table 4-7. Households who need to walk less than 5 minutes are more likely to purchase PHI policies (16.7%), this group of households account for 66.7 percent of the sample households. Although households in the group of 35-50min and group of 50-65min also have a relatively high likeliness to be policyholders, they take up a very small percentage. With respect to home's distance, a similar tendency is observed. Households located less than 1 kilometer from the town centers have a higher likelihood (21.2%) to purchase private health insurance than do other groups. In a word, households with less geographic barriers seem to have a significantly higher possibility of purchasing private health insurance policies.

Table 4-6 Physical barrier to center by policyholders and non-policyholders

	Distance from town center (km)	Walking time to highway (minutes)
Policyholder	2.8	11.8
Non-policyholder	4.4	13.7
Difference	-1.6	-1.9
Total	4.2	13.5

Source: ZEF survey 2002

Household size and household head's education

Both household size and household head's education are the leading characteristics of a household. Household size refers to the number of members who reside in the household, inclusive of the members who are migrant workers or study in cities. It is believed that a proper household size means a good structure, namely a reasonable ratio of dependants to bread earners. In the context of China, due to the family planning policy (basically one couple one child), the average household size is about 3.4 persons, according to the fifth national census in 2000 (NSB 2002). Survey results in Table 4-7 show that households with 2-4 members and 7-8 members have a higher purchase rate. The members belonging to households with 5-6 members are less inclined to purchase PHI policies for two possible reasons: first, they may have more younger dependants due to illegal births, whom no land is distributed to. Second, the couples must be fined a large amount of money for violating the family planning policy. The old dependants are somewhat not such a financial burden because their many offspring share the estovers.

Telecom tools and private health insurance purchase

Despite the dramatic development of information and communication technology (ICT) in recent years in China, Sichuan, located in underdeveloped western region, has witnessed little change in information technology access. However, 83.3% of villages in Sichuan already had access to telephone by 1997, indicating the prevalent access to telecommunication technology (Wang 2002). Telecom tools could be productive tool, consumptive tool or a mixture. The use of telecom tools can largely reduce the transaction costs and then contributes in some degree to addressing the information asymmetry in health care and insurance market.

In the questionnaire telecom tools include fixed and mobile telephones. Originally households can be divided into four groups according to the ownership of telecom tools: no telecom tool, only telephone, only mobile and both. An alternative classification to partly avoid the income effect is to treat all households with any type of telecom tools equally, i.e., only two types are grouped: with or with no telecom tools. The survey results indicate that 33.4 percent of sample households have subscribed telephone, mobiles or both. As the contingent tables in Table 4-7 shows, in terms of four-group classification, the members in a household possessing only telephone or only mobile are less likely to be insured than those in a household without private telecom tools. However, households who own both fixed and mobile phones have much higher purchase rate of private health insurance (36.1%). This phenomenon is a bit confusing given that private telecom tools diminish the transaction costs and hence facilitate the purchase of policies. A possible interpretation is that most people have equal access to public telephones; so difference in usage of telephone is not significant in spite of the difference in ownership of telephones. The fact that households possessing only mobile phone have a lower purchase rate may be associated with the high costs of mobile utilization; mobiles are still productive tools for the rural population and may be rarely used for the purpose of consumption. In terms of with-without classification, the group of households with telecom tools really has a higher purchase rate (14.2%) and the difference is statistically significant. Even then, we should be cautious about the impact of telecom tools. It remains needed to control the effect of income in the following econometric analysis.

Durables purchase and health insurance purchase

Based on the *theory of need* by Maslow (1970), the human being has five levels of need. The first level of need is physiological needs including clothes, food, shelter, transportation and other need for basic living. The second level is the need for security. The ownership of durables meets the first need, so households that have not owned these durables should generally be less likely to be insured, particularly less likely to purchase insurance policies. In the survey enumerators asked the respondents whether or not they own a television set, icebox and washer, which are traditionally regarded as the three large items of a rural household.

In the questionnaire the researcher designed four types of TV ownership: none, only black-and-white TV, only colored TV, both. As shown in Table 4-7, households with only black-and-white TV have a health insurance density of 8.1 percent, which is 2.1 percent higher than that of the group without TV set, while households with colored TV have a health insurance density two times as much as that of the group of only black-and-white TV. The difference is statistically significant ($p<0.01$). With regard to washer and refrigerator ownership, the group of households with washer or refrigerator has a significantly higher purchase rate of PHI policies. It seems that households tend to purchase durables prior to insurance. However, the poor performance of health insurance market relative to durables market should be partially responsible for this phenomenon.

Table 4-7 Cross table of household characteristics vs. PHI policy purchase

Variable		Count	Purchase rate (%)	Significance	Expected sign
Poverty status	Poor	26	10.9	0.526	+
	Average	82	13.1		
	Rich	36	10.6		
Income group	Lowest quintile	9	6.3		
	2nd quintile	18	12.5	0.001	+
	3rd quintile	26	18.1		
	4th quintile	48	33.3		
	Highest quintile	43	29.9		
Walking time	<5	96	14.6	0.012	-
	5<= and <20	18	6.2		
	20<= and <35	16	10.6		
	35<= and <50	7	14.3		
	50<= and <65	7	14.9		
	>= 65	0			
Distance from center	<1	67	21.2	0.001	-
	1<= and <3	43	8.8		
	3<= and <5	17	8.7		
	5<= and<7	2	16.7		
	7<= and<9	5	8.3		
	9<= and<11	2	2.3		
	11<	8	13.6		
Household size	2	4	14.3	0.001	-
	3	40	14.5		
	4	65	18.5		
	5	17	5.1		

	6	10	7.0	
	7	5	14.3	
	8	3	12.5	
Telecom Tools 1	Neither	80	10.2	0.003
	Only telephone	25	9.2	
	Only mobile	6	8.8	
	Both	31	36.0	
Telecom tools 2	Without	80	10.2	0.023
	With	62	14.6	
TV ownership	None	18	6.0	0.001
	Only white-black TV	31	8.1	
	Only colored TV	66	18.9	
	Both	29	15.9	
Washer ownership	Without	51	6.5	0.001
	With	93	21.6	
Refrigerator ownership	Without	107	9.6	0.001
	With	37	37.0	

Source: ZEF survey 2002

C. Health risk magnitude

Health risk of an individual is unobservable, but it affects demand for health insurance largely. So a proper proxy of health risk is employed to examine the impact of health risk on health insurance purchase. Here the researcher assumes that various households have the same likeliness to seek care after getting ill. Under this assumption the indices measuring health risk will directly affect the health insurance purchase. "Chronic disease status", and "self-rating of health status" are used to roughly measure health risk. The index of chronic disease status measures the present health risk and the index of general self-rating of health status measures the comprehensive health risk. As shown in Table 4-8, the group with chronic diseases has a relatively high density of policies. But the difference is not significant. Self-rating of health status is a rather blurry concept for respondents. However, it measures the subjective perception of health status, on which household heads make the decision of whether to purchase a policy or not. The statistic test shows an insignificant variation in purchase rate among groups of self-rating health status.

Table 4-8 Cross table between health risk indices vs. PHI policy purchase

Variable		Number of the insured	Purchase rate (%)	Significance
Chronic disease	Yes	18	13.7	0.461
	No	120	11.5	
Self-rating of health status	Very good	30	10.5	0.853
	Good	77	13.2	
	So-so	26	9.5	
	Bad or very bad	11	16.7	

Source: ZEF survey 2002

These analytical results suggest that no marked adverse selection exists in the health insurance market. The reason may be that the insurers have already excluded the high-risk groups by physical inspection. Alternatively, "health care expenses in the past year" and "times of big disease" are used in the following econometrical analyses.

D. Degree of risk aversion measured by cigarette smoking

Risk attitude is closely associated with insurance purchase. A more risk-averse agent can get a higher utility from an insurance policy than a less risk averse one does, other things held constant. However, this factor is hard to measure in practice. Propper (1989) employed a less well-defined proxy of risk attitude, whether one smokes or not. Inspired by this idea, four levels of risk aversion are defined for the research according to the numbers of cigarettes the household head smokes. Table 4-9 shows that individuals living in a household with a non-smoking household head have the highest purchase rate of health insurance, and individuals living in a household with a household head smoking most have the lowest purchase rate. Simply speaking, the more the household head smokes, the less likely a household member is to be insured.

Table 4-9 Cross tabulation of risk averse versus PHI policy purchase

Variable		Number of insurants	Purchase rate (%)	Significance
Cigarettes smoked by a household head per day	None	36	17.0	0.001
	1-15	45	13.4	
	16-30	52	12.2	
	Above 30	8	3.4	

Source: ZEF survey 2002

E. County

Table 4-10 shows a considerable variation in the purchase rate among five sampled counties. The county of Hanyuan has the highest purchase rate (29.6%) while the county of Yanyuan has a zero purchase rate. A statistical test (for both nominal variables) indicates that the relationship between "county" and "policy purchase" is highly significant ($p<0.01$). To interpret the considerable difference in purchase rate among the five counties, social and macroeconomic indicators at the county level must be combined. The lowest purchase rate in Yanyuan can be mainly explained by the far lower household's annual cash income, larger distance of home from town center and the only insurance agency, which was newly founded in county town. On the contrary, the county of Hanyuan has a relatively high cash income and a perfect insurance agency network; moreover, the agency in sampled town had an outstanding performance in sales over the past few years.

Table 4-10 Cross table of county vs. PHI policy purchase

Variable		Number of insurants	Purchase rate (%)	Significance
County	Jiange	14	5.2	0.001
	Fushun	40	16.7	
	Yanyuan	0	0	
	Hanyuan	67	29.6	
	Dongpo	23	10.3	

Source: ZEF survey 2002

4.2.3.2 Factors affecting the establishment of RCMS in a community

Community based health care schemes play an increasingly important role in the health systems of many low and middle-income countries[12]. The expectation is that community health care financing mechanisms reach population groups that government and market-based health financing arrangements have not reached. There are internationally four often-encountered forms of community financing schemes and China's rural cooperative medical scheme (RCMS) is one of the four, i.e., *community prepayment scheme* (Jakab *et al* 2001).

Where the role of governments and communities are concerned, it is worthwhile to note that the large majority of RCMS since the 1990s were set up on the community level firstly, and the risks of community were pooled again on the town level. The major role of village is in propagandizing RCMS and collecting the premium effectively and the mutual trust among villagers (village in China is equivalent to community in other countries) is crucial to the establishment of RCMS. The town governments have the major power in administrating and operating RCMS. RCMS in China can be regarded as the result of collaboration of the communities and governments, while governments, particularly the local government, play a major role. However, the governments have nothing to do with CBHIS in many other low-income countries. This is main difference between China's RCMS and other countries' CBHIS.

The prerequisite for the villagers to participate in RCMS is that a village has such a scheme, which in turn depends largely on the willingness and ability of the local villagers to pay. Only one of the five sampled counties had a RCMS in operation in 2000. The others had RCMS in traditional sense before the early 1980s. But after the disruption owing to the economic reforms, RCMS was failed to be reconstructed except a pilot RCMS. Among the ten sample villages there are two villages with RCMS. In either of the two about 95 percent of residents are the members of RCMS. The concentrated distribution of RCMS members restrained our econometric analysis on demand for RCMS at the household level. Thus, a qualitative analysis at the level of county (or village) is to be conducted as a supplement.

---

[12] In some other literatures it is documented as community based health insurance scheme (CBHIS) and in essence the two terms refers to the same thing.

The analysis suggests that RCMS is county specific. Table 4-11 shows that 34.6 percent of rural people in Dongpo district participated in RCMS, while the percentage of participants in other counties is less than 1 percent. Obviously, basic conditions of each county and characteristics of households are at least equally important, if the former is not of more importance.

A. Possible factors affecting the establishment of RCMS in a village

The researcher argues that the characteristics of the community and the county they belong to should be critical for the establishment of RCMS. In addition, institutional, organizational and managerial properties of a health insurance scheme are equally important for the sustainable development once this scheme has been set up in a village. Based on the data of the HISRS, in conjunction with the Sichuan Statistic Yearbook 2002 as well as the China dataset, a brief qualitative analysis is conducted to find out some arguments in detail.

Basic characteristics of counties

The successful establishment of RCMS in China depends to a large extent on the political and financial support by the local government. As shown in Table 4-11, all the sampled counties have constituted local official documents on how to implement RCMS. However, the vital component is how to combine the performance of implementing RCMS with the cadre's political performance and their future promotion. Dongpo district has a different incentive design from the remaining counties. Having been fully convinced of the role of RCMS in poverty prevention and poverty alleviation, the district's government took strong political measures to guarantee the implementation of RCMS. Of course, the successful establishment of RCMS is well rewarded; many of the health sector's officials as well as other officials related to the health sector have been promoted owing to the successful establishment of RCMS. This district has been set as a example of the implementation of RCMS and its experiences were widely propagandized via mass and official media and popularized to the poverty-stricken counties in western China. So the health officials have strong incentives to assist the villages and towns in establishing and developing RCMS. Despite the well-designed incentives, health insurance scheme would not have been set up and developed successfully if some basic conditions in this county had not been satisfied.

Table 4-2 demonstrates the geographic location, average net income per farmer and population density. Located in a plain, Dongpo district has better geographic and traffic conditions than the remaining sample counties, the high population density (612 persons per square km) implies that the rural people reside in a more concentrated way such that communication between households and provision of health care by health facilities are easier. Average net income per farmer in Dongpo district, 2524 Yuan, ranks second to that of Fushun county. Thus, income level per se is not a sufficient condition.

The financial support from governments is also very crucial for the establishment and the sustainable development of RCMS. The leading experience of Dongpo is that the local government is affluent sufficiently to provide financial support to the operation of RCMS. Accordingly, it is expected that Dongpo district government should spent more revenue on health care than other sample counties. Contrary to the expectation, the expenditure on health care accounted only for 2.2 percent of overall fiscal expenditure, compared with over 7 percent in both Jiange and Fushun. Obviously, the Dongpo government that subsidized RCMS spent actually less on the health sector than the others without RCMS. A hypothesis that the governments can utilize RCMS to relive them from the high cost of providing subsidies for health care is confirmed. This might be an enlightening conclusion for the poor counties that usually complain about the deficiency of fiscal revenue. The small amount of subsidies largely stimulates the initiative of establishing RCMS, which in turn mobilizes a large sum of resources into the health sector.

Table 4-11 Basic characteristics related to RCMS establishment by county

	Jiange	Fushun	Yanyuan	Hanyuan	Dongpo
Local regulation on implementing RCMS	With	With	With	With	With
Correlation to the political performance of local cadres	Loose	Loose	Loose	Loose	Close (with contract)
Experience in RCMS	Some	Some	No	Some	Much
Total fiscal spending	11170	23200	-	-	21964
Fiscal spending in health care	842	1755	-	-	475
% of total fiscal spending	7.5	7.6	-	-	2.2
Financial support from fiscal revenue	No	No	No	No	Yes
Villages covered by RCMS	14	0	0	2	190
Pop. Covered by RCMS (%)	0.7	0	0	0.1	34.6

Source: NSRHS by CASS and Statistics Yearbook of Sichuan 2002

As far as actuarial calculation of the premium and coverage is concerned, Dongpo district possesses more skills and personnel than the remaining because over ten years ago the Rand company in collaboration with Ministry of Health conducted an experiment of establishing insurance schemes in 3 villages, which lasted from 1989 to 1990. The extensive participation of community cadres and health officials in managerial activities have actually propagandized RCMS and trained many personnel for the continuous implementation of RCMS after the experiment. However, the difference in insurance knowledge and in human resources alone do not determine the establishment, given that one of three former pilot villages have ceased operating insurance schemes for many years.

To sum up, the government incentives, financial subsidies, relatively densely populated villagers, middle-income level, knowledge of insurance and human resources in management appear to constitute the necessary conditions for the successful implementation of RCMS.

Institutional, and organizational aspects of RCMS

A careful comparative analysis shows no apparent difference in organization structure of RCMS in different sample counties (see section 6.4). Nevertheless, some small difference in institutional design exists. Firstly, the top leader of the county government of Dongpo was arranged to take the full responsibility of initiating and guiding RCMS, whose mighty political power exerted a large pressure on the town governments and health officials. Another difference was the way of collecting premium. Although the cadres made efforts to persuade villagers in Dongpo district into actively joining RCMS by explaining how RCMS operates and how villagers' benefits are ensured, as a matter of fact, the premiums are collected together with the agricultural taxes. Thus, the premium collection was only partially voluntary. The cost of premium collection might become prohibitively high if the premium had to be collected on a household basis.

However, the State Council and the Ministry of Agriculture have prohibited the local governments from collecting any fees except the items of taxes and fees listed on the so called "farmer's burden card" since 1998 (Chen *et al* 2001). The officials had to take the risk of violating the regulations of the State Council if they continued to collect the premiums in this way. The immediate consequence of the execution of the regulation is that some of the towns, where rural people are not so affluent, have to forgo RCMS. The village cadres and township officials might be accused of adding farmer's burden as long as the rural households felt overloaded by taxation and some other fees. Then the cadres and town officials could be criticized by upper-level leaders or might even be dismissed. Since then, the incentives of local officials to initiate RCMS had been severely undermined; some were still in operation as the premium was hard to collect, others collapsed.

B. Reasons for rural households not to join RCMS

Two types of households are excluded from RCMS. The first group comprises those households that are reluctant to join because they are in doubt whether the insurance funds would be managed well and not be appropriated by the managers or officials for other use or even for private purpose. This kind of doubt is not groundless. The farmers observed that the insurance funds had been appropriated for such uses as constructing modern office buildings or purchasing luxurious cars for the town officials. The other type consists of those households that either have severely ill members or that are unable to afford the premium. The researcher summarized the reasons of 5 households that resided in a community with RCMS but did not join RCMS. One household did not join because it could not afford the premium. Another household was excluded because it had severely ill members; the three other households refused to join because of the distrust in the village cadres or town officials.

It is noteworthy that if over half households are unwilling to join, the move of establishing RCMS in local villages will not be approved by a villager referendum. The county office of RCMS even elevated the requirement: the enrolment rate of a village had to be no less than 80 percent to be eligible for acquiring subsidies from the county government.

4.2.4 Estimation results of demand for private health insurance

4.2.4.1 Measurement of variables and descriptive statistics

To clarify the implications of variables used in the econometric analysis, the researcher made strict definitions of them. Variable names, definitions and means of variables are shown in Table 4-12. Restricted by the data, the researcher only estimated demand for private health insurance.

Dependent variable is a dichotomous variable, 'whether or not one is insured by health insurance of company' (prihsins). Notwithstanding the different levels of premium of insurance policies, any insurance policy is treated equally. Namely, the low-premium policy for kids and students and the relatively high-premium policy for adults are considered the same. In addition, the accident insurance is viewed as health insurance because it covers the inpatient services due to accident injury.

Independent variables include individual attributes, household attributes, risk attitude, health risk magnitude and county dummies. Although many variables are theoretically correlated to the decision of purchasing private health insurance, the researcher has to elect the most important ones. The principal criterion on which the explanatory variables are selected is that they are significantly associated with the PHI purchase in the preceding descriptive analysis by the means of *contingent table* or *Pearson* correlation test. When a pair of variables has similar economic implications or has a perfect (R^2=1) or a highly close (R^2≥0.7) collinearity, the less significant one is kicked out. Another subsidiary criterion is variables' important implication for poverty reduction and rural development. Only the variables that enter the equation are defined.

As for individual attributes, only gender, age and education and 'relation to household head' are taken as explanatory variables. Two of the important indices, 'occupation' and 'marital status', are disregarded even though they are significantly related to the policy purchase in the descriptive analysis. The reason is that occupation only describes attributes of adults; it makes little sense to compare the schooling of non-adults to occupation of adults. 'Marital status' is disregarded since it is highly correlated with the variable of 'relation to household head'. On the contrary, "gender", a basic attribute of the extensive interests, is still introduced in the model even though it is insignificant in the previous table of contingency. Though degree of education is usually discrete (diploma of secondary high school or certificate of bachelor) in real society, education is still treated as a continuous variable.

As descriptively analyzed before, there are also many indices at the household level that can be utilized. Household size (HHSIZE), household head's education (HEADEDU), log cash income per capita (LG_CSHPC), log value of harvest of household (LG_HVT), home's distance from the town center (DISTANCE), and telecom tools enter the model. Income is of extensive interest but it is always difficult to single out an appropriate measurement. Aggregate household income is important and computed by summing up harvest value and cash income. The aggregate household income, however, may mask the real ability to pay. The researcher then differentiated income in cash from income in kind. To avoid the skewed distribution and over-explanation of income variables, both value of harvest and cash income per capita are transformed through a denary logarithm before estimation. 'LG_HVT' captures the information not only on income in kind but also on degree of farmers' reliance on planting agriculture in the past year. Pure planting agriculture is a very unprofitable industry nowadays in China. The researcher believes that cash income per capita is a more accurate measurement of a household's real affordability of health insurance policies. 'Distance from town center' is used to measure the geographic and physical barrier to health centers as well as agencies of insurance companies.

Housewife's education may be intimately related to policy purchase because it is believed that women are usually in charge of household financial affairs and daily expenditures. Household head's education is added because insurance purchase, after all, costs a large amount of money and must be approved by the household head. The variable of "telecom tools" is included because it can somewhat be a proxy of transaction costs. The TV set, washing machine and refrigerator display significant correlation with PHI purchase. However, they have to be excluded because of their high collinearity with cash income.

The number of cigarettes that a household head smokes is employed here to capture the information of household head's health consciousness and consumptive concept. Propper (1989) employed it as a measurement of extent of risk aversion. The researcher borrowed this idea.

Health risk magnitude is difficult to measure and it also varies across individuals. Our descriptive analysis discovers that chronic diseases, and self-rating of health status are not good measurements of health risk magnitude. But 'times of big disease' displays significant correlation with insurance purchase. Thus, the variable of health risk magnitude is measured by "times of big diseases" (BIGDISEAS).

Finally, it is expected that there are large variations that have not been interpreted by the previous variables. So the county dummies are utilized in the estimation to strengthen the power of explanation. It is worthwhile to note that the number of explanatory variables is controlled by using as few category variables as possible to be compatible with observations.

4.2.4.2 Results of probit model of demand for private health insurance

The researcher estimated the binary probit model of demand for private health insurance with different specifications based on Eq. 4-9. The parameters, standard errors and significances are presented in Table 4-13. First, in the column of probit (1) all factors except regional difference are taken into account. Because the two estimations are basically consistent, only the first estimation result is analyzed and then the slight difference between the two is discussed.

Table 4-12 Variables definition for probit model of demand for PHI

Variable Name	Definition	Probit 1	Probit 2
Dependent var.: PRIHSINS	Y=1 if one is insured by private health insurance company and 0 otherwise	Mean 0.12	Mean 0.15
Independent var.		Mean	Mean
Individual attributes			
GEND	1 If individual is male; 0 otherwise	0.509	0.511
AGE	Age in years	31.497	32.716
EDUYEARS	Formal education of individual in years	5.094	5.753
Relation to head			
REL2H_H	1 If one is head herself; 0 otherwise	0.246	0.247
REL2H_SP	1 If one is spouse of head; 0 otherwise	0.230	0.230
REL2H_KD	1 If one is kid of head; 0 otherwise	0.121	0.133
REL2H_OT	1 If one is member other than head, spouse and their kids; 0 otherwise	0.404	0.391
Household attributes			
HHSIZE	Number of household members	4.538	4.502
HEADEDU	Formal education of household head in years	5.131	5.627
DISTANCE	Distance from health care center to home (km)	3.810	3.846
LG_HVT	Denary logarithm of total value of harvest per household in 2001	3.154	3.268
LG_CSHPC	Denary logarithm of cash income per capita in 2001	3.097	3.107
COMM	1 if head is non-smoker; 0 otherwise	0.345	0.255
Risk aversion			
CIGAR_NO	1 if head does not smoke; 0 otherwise	0.178	0.177
CIGAR_MN	1 if head smokes below 1 packet a day; 0 otherwise	0.187	0.163
CIGAR_TM	1 if head smokes over 1 packet a day; 0 otherwise	0.650	0.492
Health risk magnitude			
BIGDISEA	Big disease times since establishment of home	2.038	2.063
County dummies			
CUNTY_JG	1 If one lives in County Jiange; 0 otherwise		0.220
CUNTY_FS	1 If one lives in County Fushun; 0 otherwise		0.238
CUNTY_HY	1 If one lives in County Hanyuan; 0 otherwise		0.216
CUNTY_DP	1 If one lives in County Dongpo; 0 otherwise		0.257
No of observation		968	837

Note: the sample in Probit (2) differs from that of Probit (1) by excluding the data of Yanyuan, where the purchase of PHI is perfectly predicted to be zero.

Estimation results show that the *individual characteristics* except gender affect demand for PHI significantly. Their coefficients also bear expected signs. The coefficient of age indicates its slightly negative impact on PHI policy purchase. The major reason lies in the fact that a large majority of kids are insured by pupil health and accident insurance. Another possible reason is that the one-child families give priority to insuring the only child. Individuals' education seems to affect the health insurance purchase highly significantly. The more one is educated, the more likely one is to be insured. The relation to household head has a significant impact on policy purchase. Compared with the housewives, household head, kids and other groups (mainly the parents of household heads) are all more likely to be insured. Moreover, the kids and the old men tend to be insured while household heads and housewives are less likely to be insured. In short, the significance of most individual characteristics demonstrates that the private health insurance is individual specific.

Among the individual characteristics concerned, age, gender, and education are suggested to be the most important explanatory variables. Henderson *et al* (1994) used a logit model to estimate demand for health insurance in China (in broad sense, including private, social health insurances and RCMS in both urban and rural areas). Regarding age both their result and that of this study show that the group at the age of 0-14 has the highest likeliness to be insured. However, in their research the old group (above 60) had the second highest rate of participation while in our study the older are least likely to be insured. The coefficient of age in the present study is –0.019, compared with the coefficient of age in Henderson's (1998) probit estimation (0.024). The similar absolute values have opposite directions. Costa *et al*'s (1999) findings also exhibited that demand for PHI rose with age. Even so, our result is reliable because the rural parents attach much importance to the children and the insurance policies for children are much cheaper than those for adults. An overwhelming majority of the children's health insurance is bought on the basis of school that children are attending. In Henderson's two researches in China, education did not show significant impact on insurance participation while gender did. Females were significantly less likely to be insured. The present research, however, shows a discrimination against males but it is not statistically significant. In rural areas, the relation to household head is also a good explanatory variable. The blood relations to household heads seem to determine that a certain type of members are more likely to be insured, e.g., kids[13].

Among six variables describing household's attributes, two of them affect demand for PHI significantly and all the parameters bear expected signs. It is noteworthy that, contrary to the expectation, harvest value, as a measurement of income in kind, affects the purchase significantly negatively. As mentioned in preceding descriptive analysis, this may be due to the underdeveloped food grain market, in which people cannot trade the food grains (major segment of harvest) at a reasonable price to pay for insurance policies. Another possible reason is that a large share of harvest value in total household income means deep involvement in the grain food production, which is less profitable in China today.

---

[13] However, waters (2000) show that the spouse is more likely to be insured in the case of Ecuador.

Household head's education, home's distance from center, and telecom tools do not affect the purchase of PHI significantly, but they bear expected signs. The negative sign of household size can be understood by the fact that a big household has a relatively strong ability to cope with risk and hence retains part of health risk it faces.

The existing researches that used urban areas as main study areas did not pay due attention to household characteristics except income and geographic barrier (e.g., Henderson *et al* 1998; Propper 1989; Costa *et al* 1999). Sufficient attention is paid to household variables in the current study, but only home's distance from the center shows a significant influence on PHI purchase after the regional difference is controlled. In an innovative way income is classified into income in cash and income in kind. The significantly negative impact of income in kind suggests that food grains have a great barrier to market. The deeper rural households are involved in food grain production, the less cash income they earn and the less likely they are to purchase insurance policies. Henderson *et al* (1994) demonstrated similar results, but his results were based on a comparison of farmers with other occupations.

Where risk aversion is concerned, household heads that smoke over one packet of cigarettes per day are significantly less likely to buy insurance policies for any household member than non-smoking household heads. However, moderate smokers are not significantly different from non-smokers. The difference in PHI policy purchase lies in the belief that non-smokers pay more attention to their own and other household members' health, that is, they are more risk averse than strong smokers.

A range of approaches of measuring risk aversion was introduced in insurance research. Manning *et al* (1996) directly estimated the revealed risk aversion parameter using experimental data on insurance preference over full coverage and catastrophic limit insurance. Costa *et al* (1999) used a qualitative variable ranging from 1 (extremely risk averse) to 10 (risk lover) to measure risk attitude. We used a less well-defined approach—cigarettes that household head smokes everyday to measure the extent of risk aversion, following the idea of Proper (1989). In Costa's estimation the coefficient of risk attitude bears the expected sign just as in the present research.

The magnitude of health risk measured by 'times of big diseases' since birth bears a expected sign and affects the health insurance purchase significantly, implying that adverse selection exists despite the strict screening. Unlike this result, in Costa's research, risk magnitude did not show a significant correlation with the number of outpatient visits in the past three months. However, a problem arises when the health risk magnitude is measured because the relation between health insurance purchase and health status is not linear (Propper 1989). That is, the individuals in very good health (low health risk magnitude) and poor health are least likely to purchase, because the healthy can benefit little from the policy while the unhealthy, especially those with some major disease, are declined to purchase owing to the insurer's screening. Thus, the relation of health risk magnitude with insurance purchase seems inconsistent among different researches.

Secondly, the county dummies are introduced in specification (2). Not surprisingly, the model's capacity of explanation rises as these dummies contain abundant information that the previous ones have not captured, such as local infrastructure, extent of trust in insurer, development degree of health insurance market and relevant policies. Note that the county of Yanyuan has been excluded because the sub-data can perfectly predict the dependent variable to be zero (not to purchase). Using Jiange as a reference county, Hanyuan, Fushun and Dongpo's households are more likely to buy health insurance policies. The coefficient of Hanyuan county is bigger than others, because its rural agency network of insurance company was well developed and the insurance agents worked more efficiently. One of Hanyuan company's agencies ranked among the top ten in sales of insurance policies in Sichuan province and over 60 percent of sales were health insurance policies in 2000.

It is worthy to note that all significant coefficients in probit (2) still bear the same signs as probit (1) and change slightly in magnitude, only the coefficient of cash income per capita decreases by about 35 percent. The slight change shows the high stability of the model.

It is also noticeable that the coefficient of distance from town center turns from insignificant to significant. The change is reasonable, as the geographic barriers may discourage people from seeking care and hence from purchasing health insurance. In addition, CIGAR_TM, a significant variable in probit (1) becomes insignificant in probit (2) though they still bear the same sign. This change may illustrate that the behavior of heavy smoking does not significantly hinder the smokers from purchasing insurance policies or it is not a good proxy of risk attitude.

Marginal effect and income elasticity

So far only the significance and directions of effects of independent variables on private health insurance purchase have been explored. The magnitudes of impacts measured by marginal effect and elasticity are of more interest. However, unlike linear regression equations estimated through OLS, the parameters of a non-linear equation are not necessarily marginal effects. Based on the approach elaborated on by Greene (2000), the marginal effect is computed in the following way.

Eq. 4-16 $$\frac{\partial E[y|x]}{\partial X} = \left\{ \frac{dF(\beta X)}{d(\beta X)} \right\} \beta = f(\beta X)\beta$$

where *f(.)* is the density function that corresponds to the cumulative distribution, *F(.)*. For the normal distribution, this result is

Table 4-13 Estimation results of purchase of PHI: Binary probit model

Dep. variable: PHI purchase=1	Probit (1)		Probit (2)	
Indep. variables	Coefficient	Standard Error	Coefficient	Standard Error
Individual attributes				
GEND (1=male)	-0.0296	0.1440	0.0735	0.1538
AGE	-0.0141***	0.0052	-0.0119**	0.0052
EDUYEARS	0.0717***	0.0190	0.0617***	0.0203
Relation to head (reference= spouse)				
REL2H_H	0.5750**	0.2688	0.5170*	0.2783
REL2H_OT	0.9609***	0.3080	0.9795***	0.3212
REL2H_KD	0.8938***	0.2524	0.9238***	0.2592
Household attributes				
HHSIZE	-0.0157	0.0537	-0.0241	0.0599
HEADEDU	0.0236	0.0209	0.0243	0.0221
DISTANCE	-0.0197	0.0163	-0.0606***	0.0215
LG_CSHPC	0.8298***	0.1829	0.4696**	0.2153
LG_HVT	-0.4961***	0.1377	-0.4885***	0.1430
COMM	0.0211	0.1346	0.0113	0.1575
Risk aversion (Reference= non-smoker)				
CIGAR_TM	-0.6169***	0.2337	-0.3355	0.2402
CIGAR_MD	-0.0755	0.1589	0.1635	0.1408
Health risk magnitude				
BIGDISEA	0.0787**	0.0355	0.0739**	0.0360
County dummies (Reference=Jiange)				
CUNTY_FS			0.7505***	0.2244
CUNTY_HY			1.2598***	0.2314
CUNTY_DP			0.4252***	0.2250
Constant	-3.1420***	0.7684	-2.6620***	0.9232
Number of observations	968		837	
Log likelihood	-271.63		-250.03	
Chi-squared	146.281		155.03	
Significance level	0.001		0.001	

Source: computed from ZEF survey 2002 Note: *p<0.10, **p<0.05 and ***p<0.01.

Eq. 4-17 $$\frac{\partial E[y|x]}{\partial X} = \phi(\beta X)\beta$$

where *ϕ(.)* is the standard normal density. It is obvious that these values vary with *x*. Marginal effect is normally calculated at the mean of regressor and is applied to the slope of the entire model. Another complication for computing marginal effect in a binary choice model arises because regressors usually include dummy variables. Since the derivative is with respect to a small change, Eq. 4-16 is no longer applicable to compute marginal effects of dummy variables. The applicable formula would be

Eq. 4-18 *Marginal effect*= $\Pr ob(Y = 1 | \bar{x}_{(d)}, d = 1) - \Pr ob(Y = 1 | \bar{x}_{(d)}, d = 0)$

where $\bar{x}_{(d)}$ denotes the means of all the other variables in the model. The marginal effects of the significant coefficients in two probit models are presented in Table 4-14. Now a change in any one of the independent variables exerts a direct impact on the probability of PHI purchase. For instance, a one-year-old increase around the mean of age leads to a decrease in probability of buying health insurance by 0.17 percent. Likewise, a one-year increase in formal education around the mean of educational years raises the likelihood of purchasing health insurance by 0.85 percent. Compared with those of other covariate variables, the marginal effects of the two variables are weak.

It is interesting that the dummy variable 'relation to head' has a strong marginal effect on health insurance. For example, kids are 10.6 percent more likely to be insured by private health insurance than non-core members of the household. With respect to home's distance from town center, a one-kilometer increase in distance results in a decrease in the probability of being insured for all household members by 0.23 percent. The marginal effect of income deserves specific attention but is neglected here, as it will be explained in detail in terms of elasticiy.

As far as the regional difference is concerned, the probability of being insured for people in Hanyuan county is 16.6 percent higher than that of Jiange county (reference group). By the same token, people in Dongpo and Fushun are 9.9 and 5.6 percent more likely to be insured by a private health insurance policy.

Table 4-14 Marginal effects on probability of PHI purchase

	Probit (1)		Probit (2)	
Variable	Coefficient	Standard Error	Coefficient	Standard Error
AGE	-0.0017***	0.0006	-0.0016**	0.0007
EDUYEARS	0.0085***	0.0022	0.0081***	0.0027
REL2H_H	0.0680**	0.0312	0.0682*	0.0361
REL2H_OT	0.1137***	0.0351	0.1293***	0.0410
REL2H_KD	0.1057***	0.0294	0.1219***	0.0336
DISTANCE	-0.0023	0.0019	-0.0080***	0.0027
LG_CSHPC	0.0981***	0.0215	0.0620**	0.0288
LG_HVT	-0.0587***	0.0169	-0.0645***	0.0195
CIGAR_TM	-0.0730***	0.0276	-0.0443	0.0316
BIGDISEA	0.0093***	0.0042	0.0098**	0.0048
CUNTY_FS			0.0991***	0.0295
CUNTY_HY			0.1663***	0.0304
CUNTY_DP			0.0561*	0.0294

Source: computed from ZEF survey 2002 Note: *p<0.10, **p<0.05 and ***p<0.01.

Despite the usefulness of marginal effects, many health economists are used to adopting the elasticity to assess impacts of income and price (e.g., Holmer 1984, Marquis *et al* 1987), since elasticity is normalized for the variables' units. Elasticity is defined as the percentage change in one variable that is associated with a one-percent change in another variable. In order to compare our results with the previous ones, an attempt is also made to compute the elasticity. Following the approach elaborated on by Train (1993), the elasticity of probability P with respect to x, a variable entering the utility of alternative i, is

$$\text{Eq. 4-19} \quad E_{xi} = \frac{\partial P_{ni} / P_{ni}}{\partial X_{ni} / X_{ni}} = \frac{\partial P_{ni}}{\partial X_{ni}} \frac{X_{ni}}{P_{ni}} = \frac{\partial V_{ni}}{\partial X_{ni}} P_{ni}(1-P_{ni}) \frac{X_{ni}}{P_{ni}}$$

$$= \frac{\partial V_{ni}}{\partial X_{ni}} X_{ni}(1-P_{ni})$$

where P_{ni} is the probability of individual i selecting choice J_n and X_{ni} is the independent variable. The utility of individual i obtaining from alternative n, denoted U_{ni}, can be decomposed into a part that is known by the researcher, labeled as V_{ni} and an unknown part that is assumed to be a random variable, labeled as e_{ni}. This is expressed $U_{ni}= V_{ni}+ e_{ni}$. We turn to the normal distribution of e_{ni}.

$$E = \frac{f(\beta X).\beta.X}{F(\beta X)} = \frac{\phi(\beta X).\beta.X}{\Phi(\beta X)} = ME * \frac{X}{\Phi(\beta X)}$$

where ME is marginal effect and $\Phi(.)$ is the cumulative function of standard normal density of distribution. Then we can use the derivatives to compute the elasticity by letting the X be mean and $\Phi(.)$ be the cumulative probability when X is taken mean.

The elasticities of cash income per capita and harvest value per capita for specification (1) are 0.313 and -0.191, respectively. In most studies using individual data, estimated income elasticity is generally positive, but less than 1 for almost any measure of insurance chosen (Phelps 1997). Thus, the elasticity of cash income per capita is consistent with previous estimated results using individual data. Inconsistent with previous elasticities, the harvest value per capita, part of the total income for most rural household, has a negative elasticity, but its absolute value still falls into the range from zero to one.

4.3 Impact of health insurance on health care-seeking behavior

As stated previously, health care is the weightiest health input; none of other health inputs has the full substitution for health care. When the stock of health declines sharply, or the rate of depreciation rises rapidly, only health intervention with appropriate health care can guarantee the stock of health to be above H^*, a threshold of health to be alive. However, the quantity of health care demanded is affected by its price, which can be considerably reduced if the ill are insured. The impact of health insurance on quantity of health care demanded used to be neglected in rural areas because there were few rural Chinese who are insured. However, the figure has been in rise in last decade. It was reported that about 10 percent of rural Chinese had part of their health expenses paid by a third party—an insurance company, social security scheme or rural cooperative medical schemes (MOH 1999). There is, therefore, an increasing need to analyze the impact of health insurance on health care seeking behavior of rural Chinese.

4.3.1 Modeling health care seeking behavior with health insurance

4.3.1.1 The behavioral model of seeking care

To assess the impact of health insurance, it is required to model health care seeking behavior and estimate of the parameters of behavior function, simultaneously taking into account health insurance status of individuals. Price elasticity, which is computed from the estimated model, is especially useful, indicating how user fees affect patients' utilizations and providers' revenues.

Decision making on care seeking behavior is a complicated process. Some studies look on decision making as a three-stage process: reporting illness, making decision of seeking treatment and making choice of providers (Chawla *et al* 2000; Asfaw 2002). In this thesis, it is assumed that household head makes decision only when a household member reports sickness. The forthcoming decision procedure includes three steps. First, a household head makes a decision of whether to seek treatment or not. The expected benefit from consuming health care is an improvement in health, and the cost is the reduced consumption of other goods and services. Second, the household head makes up his mind in which health facility the ill should seek care. Each provider offers an expected improvement in health (efficacy) for a price. There may be a third step, in which household head decides whether the ill should move to another type of health facility if the disease can not be cured or the household can not afford the costs any more. A decision tree is helpful to illustrate this procedure (Figure 4-7). However, only the first two steps of care-seeking behavior (decision of seeking treatment and choice of the first providers) are investigated and econometrically analyzed in the study. The third step is neglected. Each step of this decision may depend on a vector of characteristics of the ill, a vector of characteristics of the household, health insurance status, and characteristics of providers.

Figure 4-7 Two-stage decision tree: seeking care and choice of providers

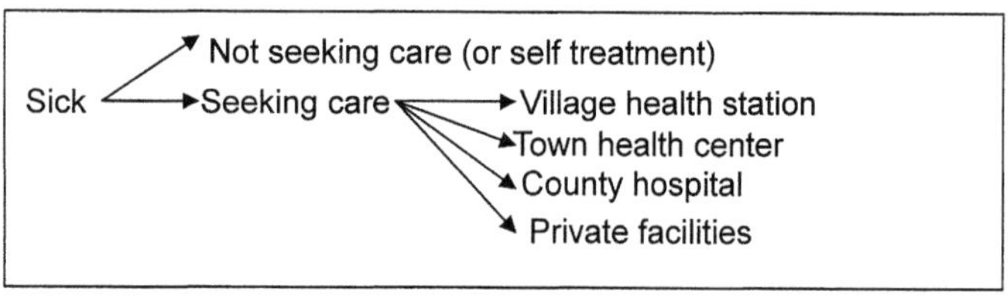

The quality issue should not be overlooked in the study of care seeking behavior. In African countries the quality of medical care is well recognized as a key factor in determining the success of health care financing reforms (Mwabu 1990 and literatures cited therein). In China, Hsiao (1995) observed that income elasticity to quality of medical care appears quite high. Medical care quality assesses how well medical care produces health improvement. Thus, the quality has bearing on whether the medical intervention was appropriately selected and properly carried out. Though assessment of medical care quality is problematic both to consumers and to health care providers, patients can simply assess the quality according to their experiences and/or attributes of health facilities. One common indicator of quality is the number of trained doctors a facility owns.

China's three-tier health care system is a carefully designed three levels of facilities and can be a good object of research on demand for quality. As shown in Table 4-15, the difference in quality of village health station, town health center and county hospital is evident and the quality escalates tier after tier. However, the quality of private clinics is ambiguous. A very small portion of private clinics is comparable to county hospital, while a large majority of the private clinics are better than village health stations but worse than town health centers. Thus, the choice of providers is to a large extent the selection of quality.

Table 4-15 Difference in quality of alternative providers

Health Facilities	Number of hospital beds	Number of staff	Quality description of major professionals
County hospital	126	180.4	Physicians of 4-5 years of higher medical education, nurses, technicians and pharmacists. Being able to provide almost all types of out- and inpatient services, operations, tests as well as western and Chinese medicines.
Township health center	15	24.3	Physicians of 3 years of medical school education after high school, assistant physicians and village doctors.
Village health station	n.a.	1.8	Village doctors who are trained for 3- 6 months after junior high school and received on average 2-3 weeks of continuing education each year.
Private clinics	n.a.	1.4	Physicians of 3 years of medical school education with independent license issued by county health bureau, and nurses.

Source: NSB 2002 and Dong 2000

Let the expected utility conditional on receiving care from provider j be given by equation 4-1, which is rewritten here for convenience and add the subscript.

Eq. 4-20 $$U_j = U(H_j, C_j)$$

where H_j is the expected health status after receiving treatment from provider j and C_j is consumption net of costs of obtaining care from provider j and premium for the health insurance (of course, the premium is zero if the ill does not purchase health insurance). Health production is modeled as a relation between the health outcome and a set of health input choices; its shape depends on the underlying health technology.

Eq. 4-21 $H_j = H(m_j, C_j; X)$

where X is a vector of exogenous characteristics influencing individuals' health and m is health care consumed. Individuals will maximize their expected utility subject to budget constrain

Eq. 4-22 $C_j + P_j^* \leq Y$

where P_j* is the total price of providers j's care. The total price of health care comprises direct payment (user fees charged by provider) and indirect costs of access (travel costs plus waiting time and nursing time by other household members). Note that user fees are only a proportion of the total health care price, αP_j when the individual is insured. α is coinsurance rate of policy and P_j is the charge of provider j, because insurance company will reimburse $(1-\alpha)P$. Substitution of Eq. 4-22 into Eq. 4-20 yields the conditional indirect utility function

$$U_j = U(H_j, Y - P_j^*)$$

Notice that income influences utility through the consumption term and that health insurance impacts demand for health care through price of health care. Then faced with j+1 provider alternatives (the j=0 alternative being self-care). The unconditional utility maximization problem is

Eq. 4-23 $U^* = \max(U_0, U_1, ..., U_j)$

where U* is the maximum utility. The solution to Eq. 4-23 gives the alternative that is chosen and, where there are random terms in the model, the probability of each alternative is chosen.

4.3.1.2 Empirical specification

The solution to Eq. 4-23 yields a system of demand functions whose forms are probabilities of the alternatives being chosen. The probability of a particular alternative being chosen equals the probability of this choice yielding the highest utility among all the alternatives. In a discrete modeling framework, this probability is interpreted as the demand function; its functional form depends on the functional form of the conditional utility function and the distribution of the stochastic terms. Assume that the utility is taken a linear approximation in health and consumption, which implies the conditional utility function Eq. 4-20 is now

Eq. 4-24 $U_j = a_{1j} H_j + a_{2j} C_j + \varepsilon_j$

where ε_j is a zero mean random disturbance with finite variance and is uncorrelated across individuals and alternatives, and a_{1j} and a_{2j} are parameters to be estimated. This error term is assumed to have a Gumbel distribution, leading to the multinomial logit specification. Household heads make choices based on the comparison of indirect utility function for each variety of health care available, including that of self-treatment. However, we have to redefine the budget constraint when the impact of insurance has to be taken into consideration. Suppose for the simplicity that health insurance policy is characterized by a premium π, and a coinsurance rate α. The simple health insurance policy is described as (π, α) without loss of generality (the attributes of deductible and maximum claims are ignored). Now the insured has only to pay αP_j at a cost of π. The budget constraint denotes now

$$C_j + \alpha P_j = Y - \pi$$

Note that the uninsured individuals can also be modeled by conveniently letting both π and α be zero. Rearranging the equation yields

$$C_j = Y - \pi - \alpha P_j$$

adding the indirect opportunity cost of time to the equation yields

Eq. 4-25 $$C_j = Y - \pi - \alpha P_j - wT_j$$

where w is average wage of time and T_j is the sum of waiting time, accompanying time and nursing time. Substitution of Eq. 4-25 into Eq. 4-20 yields

$$U_j = U(H_j, (Y - \pi) - (\alpha P_j^* + wT_j))$$

Specifically, the concrete linear form of utility function can be written

Eq. 4-26 $$U_j = a_{1j} H(m_j, C_j; X) + a_{2j}[(Y - \pi) - (\alpha P_j^* + wT_j)] + \varepsilon_j$$

There remains a possibility of an ill policyholder selecting self-treatment owing to the presence of co-payment. Even without co-payment, treatment is not completely free because the ill has to spend time and money in traveling and waiting for treatment. As $P_0 = H_0 = 0$, the conditional utility function for the self-care alternative is

Eq. 4-27 $$U_0 = a_{10} H(m_0, C_0; X) + a_{20}(Y - \pi) + \varepsilon_0$$

To illustrate the role of price, income and health insurance status, we normalize the utility from the self-treatment alternative to zero by subtracting Eq. 4-27 from Eq. 4-26. In this case Eq. 4-24 becomes

Eq.4-28

$$U_j - U_0 = a_{1j}(m_j, C_j; X) - a_{10}(m_j, C_j; X) + a_{2j}(Y - \pi) - a_{20}(Y - \pi) - a_{2j}(\alpha P_j^* + wT_j) + \varepsilon_j - \varepsilon_0$$

This is the structural demand equation. The option giving the highest $(Uj - U_0)$ is the one chosen. If the reduced form of health production function is also taken linear approximation

Eq. 4-29 $$H_j = \beta_{0j} + \beta_{1j} X + \beta_{2j} M + \beta_{3j} C + \eta_j$$

where η_j is zero mean random disturbance with finite variance. Substituting Eq. 4-21 and Eq. 4-29 into the conditional utility functions in Eq. 4-26, which appears in all of the conditional utility functions, gives

Eq. 4-30

$$U_j = \beta_{0j} + \beta_{1j}X + \beta_{2j}M + \beta_{3j}C + a_{2j}(Y-\pi) - \alpha_{20}(Y-\pi) - \alpha_{2j}(\alpha P_j^* + wT_j) + \eta_j + \varepsilon_j$$

The reduced form of Eq. 4-30 can be written as a function of estimable parameters by ruling out some repetitions.

Eq.4-31

$$V_j = \alpha_{10} + \beta_{10}A_i + \gamma_{10}E_i + \delta_{10}S_i + \varphi_{10}D_i + \eta_{10}SE_i + \tau_{10}ET_i + \kappa_{10}CI_i + \lambda_{10}WT_{ij} + \theta_{10}CH_{ij} + \vartheta_{10}HI_i + \varepsilon_{10}$$

Where A_i= years old of individual *i* who selects choice *j*

E_j = years of formal education of individual *i*

S_i = sex dummy of individual *i* (1 if male, 0 if female)

D_i = duration of illness of individual *i*

SE_i = years of spouse's formal education

ET_j= ethnic group (1=minority, 0 otherwise)

DIS_j= distance from home to town center

CI_{ij} = logarithm of per capita cash income

WT_{ij} = logarithm of time spent on treatment and nursing *j*

CH_{ij} = logarithm of user fees charged by health facilities *j*

HI_i = private health insurance purchase of individual *i* (1 if policyholder, 0 otherwise)

4.3.1.3 Estimation techniques

A binary logit model is used to simulate decision of seeking treatment and then is estimated. A multiple-choice model is used to simulate the choice of the providers because there are more than two providers for the ill to choose. Two commonly used multiple-choice models are multinomial logit model (MNL), and the McFadden logit model. However, the latter demands characteristics data on each choice, which is unavailable in our dataset, so the multinomial logit model is selected. The dependent variables can be many types in accordance with the correlation between the choices: unordered, sequential, and ordered categories. In this section, the choices are looked on as the unordered type because a strict referral system in rural China has not existed anymore since the rural cooperative medical system broke down in the early 1980s. The ill people can visit the best-equipped county hospital directly as long as they think it necessary to do that.

Suppose there are *m* categories of health care providers. Let $P_1P_2...P_m$ be the probabilities associated with these *m* providers. Then the multinomial logit model is equivalent to expressing these probabilities in a binary form. Let

$$\frac{P_1}{P_1 + Pm} = F(\beta_1' x)$$

$$\frac{P_2}{P_2 + Pm} = F(\beta_2' x)$$

$$\frac{P_{m-1}}{P_{m-1} + Pm} = F(\beta_{m-1}' x)$$

These imply

Eq. 4-32 $$\frac{P_j}{P_m} = \frac{F(\beta_j' x)}{1 - F(\beta_j' x)} = G(\beta_j' x) \qquad (j=1,2,\dots,m-1)$$

Because

Eq. 4-33 $$\sum_{j=1}^{m-1} \frac{P_j}{P_m} = \frac{1 - P_m}{P_m} = \frac{1}{P_m} - 1$$

We have

Eq. 4-34 $$P_m = \left[1 + \sum_{j=1}^{m-1} G(\beta_j' x)\right]$$

And hence

Eq. 4-35 $$P_j = \frac{G(\beta_j' x)}{1 + \sum_{j=1}^{m-1} G(\beta_j' x)}$$

Although any of the previous mentioned underlying distributions of u could in principle be used, from the computational and explanatory viewpoint the logistic is easiest to handle. In this case $G(\beta_j' x)$ in Eq. 4-34 is $\exp(\beta_j' x)$, and the Eq. 4-34 and Eq. 4-35 can be written as

$$P_j = e^{\beta_j'} / D \qquad (j=1,2,\dots,m-1)$$

and $P_m = 1/D$

where $D=1+\sum_{j=1}^{m-1} e^{\beta_j' x}$

This model is commonly referred to as multinomial logit model and the method of maximum likelihood is applied to estimating. The estimating techniques can be seen in the textbook of econometric analysis (Green 2000).

4.3.2 Descriptive analysis on health care seeking behavior

4.3.2.1 Illness report

A basic assumption is that those who do not report illness do not demand health care. This section examines only health care utilization of people who report illness. In the survey three indices with respect to illness report are defined to reflect the incidence of illness: incidence of illness in the past month, incidence of illness in the past year, and whether a chronic disease is reported. As shown in Figure 4-8, there were 282 of 1217 sampled persons, or 23.2 percent who reported illness in 2001; 10.3 and 10.2 percent of sampled persons reported illness in December and chronic disease in 2001, respectively. However, the coming analysis focuses on the incidence of illness in 2001. Additionally, the severity of illness measured by duration of illness is combined with incidence of illness in 2001 because the respondents appear to neglect slight illnesses. The average duration of disease reported is some 61 days (those who did not report any disease are excluded from the mean calculation). Before investigating health care seeking behavior of people who reported illness, the possible reasons for the large variation of frequency and duration of illness reports are to be identified.

Figure 4-8 Illness reports in 2001, in past month and chronic disease report

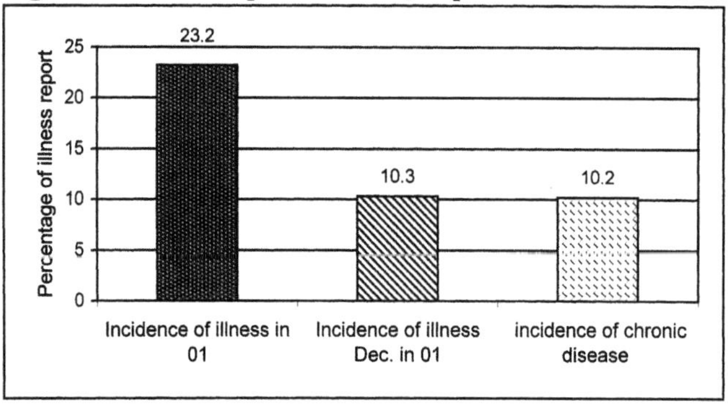

Illness report vs. income

There exist two opposite findings on rate of illness report with respect to income. Some studies found a higher rate of illness report in lower income groups (e.g., Haynes 1999), whilst others found a lower one in lower income groups (e.g., Thompson *et al* 2001). Our finding is in support of the latter.

To identify whether the poor are more likely to report illness or are ill longer than the non-poor, the households' total income (in kind plus in cash) is grouped into 5 quintiles. As shown in Figure 4-9, the rate of sickness report grows from 15.6 percent for the lowest quintile to 25.9 percent for the highest quintile. However, the duration of illness alters in diametrically opposite directions. The poorest quintile has the longest duration of illness while the richest have the shortest duration. The oppositely altering directions of illness report and illness duration suggest that the poor people do not perceive illness sensitively, or they only think of a relatively severe discomfort as illness, e.g., a discomfort that forces them to stop working or schooling. Under this assumption, the phenomenon of high illness duration but low illness report is interpretable. Not surprisingly, the difference between groups is statistically significant.

Illness report vs. demographic characteristics

Various demographic characteristics are also expected to be associated with illness report or duration of illness. Table 4-16 summarizes major demographic characteristics intimately related to illness report. Females are normally thought to have a higher morbidity than males. The present result supports this a priori expectation, though the difference between males and females is not statistically significant. There is also a slight and insignificant difference in duration of illness. With the increase in age people become more liable to diseases and the diseases last longer. Moreover, it is outstanding that the above-60 group has a much higher morbidity and longer duration of disease. The large number of illness days (147 days) suggests that most of them suffer a chronic disease.

Figure 4-9 Incidence and duration of illness vs. household total income

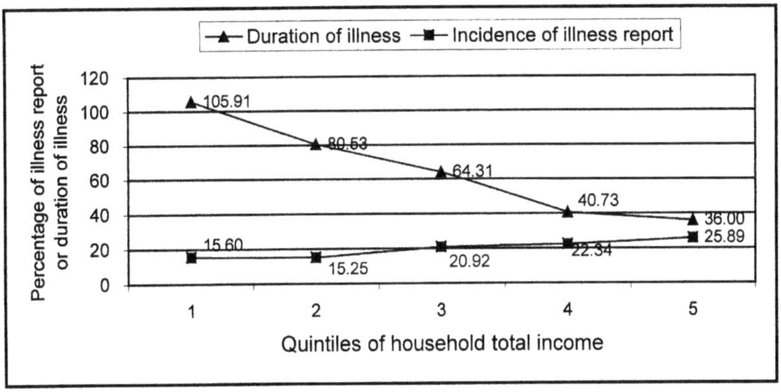

Education has an impact on the efficiency of health production. The more educated one is, the more efficiently one produces health with consumption and health care and hence the less likely one falls ill. The findings clearly confirm this hypothesis. Considering that a higher education may simultaneously increase consciousness of illness, the *de facto* role of education in improving health tends to be underestimated. The more educated groups also have a shorter duration of illness, further verifying the significantly and intimately positive relationship between health and education.

An inspiring finding is that the vaccination decreases morbidity by 7 percent and duration of illness by 60 days by comparing the vaccinated with the unvaccinated group. In contrast, people with the chronic diseases have a much longer duration of disease and a much higher morbidity than those without chronic diseases. The exploration on self-rating health status versus illness report may offer an edificatory idea on how people assess their own health status. The findings demonstrate an intimate correlation between perceptions of health status and illness report ($p<0.01$), namely, the more frequently one falls ill, the worse one assesses his or her health status, *vice versa*.

Table 4-16 Illness report vs. demographic characteristics

Demographic factors		Rate of illness report			Duration of illness*		
		N	Mean	Significance	N	Mean	Significance
Gender	Male	139	23.4	0.829	139	62.1	0.828
	Female	143	22.9		143	59.3	
Age	0-14	43	15.9	0.001	43	30.7	
	15-29	62	18.7		62	40.5	0.001
	30-44	78	24.8		78	34.3	
	45-59	61	28.5		61	84.3	
	Above 60	37	44.6		37	147.3	
Education	No education	90	32.3	0.013	90	84.2	0.015
	Primary education	69	24.7		69	68.4	
	Junior school	42	15.1		42	31.0	
	Senior school	56	20.1		56	51.7	
	College	22	7.9		22	18.7	
Vaccination status	Vaccinated	231	22.2	0.043	231	49.9	0.001
	Unvaccinated	51	29.1		51	109.6	
Chronic disease	With	79	63.7	0.001	79	114.9	0.001
	Without	201	19.0		201	40.0	
Health status by self-rating	Very good	26	9.1	0.001	26	33.9	0.001
	Good	98	16.8		98	26.6	
	So-so	103	37.6		103	37.4	
	Bad and very bad	55	80.9		55	177.8	

Source: ZEF survey 2002

Note: *mean of duration of illness is calculated only for members reported illness or injury.

Illness reports vs. health insurance status and county

Both morbidity and duration of illness vary largely from members to nonmembers of RCMS, but there is no difference between policyholders and non-policyholders (Table 4-17). When comparing members with non-members of RCMS, the researcher finds out that members have a much higher morbidity (41.5%) than non-members (19.6%). Meanwhile, the difference is of highly statistical significance ($p<0.01$). It is expected that the purchase of PHI policy have a similar effect on the illness report. Although policyholders of PHI have a slightly higher morbidity than non-policyholders, the difference is not statistically significant.

The incidence of illness varies substantially across counties. A large proportion of variation can be explained by the factors identified above. The highest incidence of illness in Dongpo is about 4 times as large as the lowest in Jiange (Table 4-18). Given the large variation in household's total income across counties, this is not surprising. Except the minority county of Yanyuan, the county of Jiange with the lowest average total income per household (10318 Yuan) has the lowest illness reports, whereas the county of Dongpo with the highest average total income per household (17055 Yuan) has the highest (39.9%). Thus, households in poor counties tend to report the serious illness only. Furthermore, morbidity data in Sichuan extracted from NRSHS (CASS 2000) offers a consistent result. Illness report in the past two weeks priori to the survey varies considerably across 6 sampled counties in Sichuan, from 2.8% to 38.5%.

Table 4-17 Illness reports in 2001 by county, health insurance status

		Incidence of illness			Duration of disease (day)		
		N	%	Significance	N	Mean	Significance
RCMS	Non-member	199	19.60	0.001	199	70.61	0.013
	Member	83	41.50		83	36.86	
Private health insurance	Policyholder	247	23.00	0.731	247	59.49	0.613
	Non-policyholder	35	24.30		35	69.06	
County	Jiange	28	10.40	0.001	28	187.64	0.001
	Fushun	40	16.60		40	51.13	
	Yanyuan	53	20.60		53	66.32	
	Hanyuan	72	31.60		72	44.51	
	Dongpo	89	39.90		89	34.74	

Source: ZEF survey 2002

The duration of disease varies substantially across counties too. In Jiange, the mean of duration of disease averages 187.6 days, which is 5.4 times as much as that of Dongpo district. However, incidence of illness reported in Jiange is much smaller than that in other counties. At the same time, enormous variation in illness report arouses a suspicion of the soundness of examining health care seeking behavior among the ill. Quite a few people in counties of low incidence of illness probably did not report the slight sickness for which they had not sought care. They may neither have perceived the sickness or feel kind of self-contemptuous of bringing out the fact that they cannot afford the costs of health care. Another preliminary conclusion we can draw is that the counties with a lower incidence of illness have a higher severity of illness.

The highest illness report in Dongpo county could be partially explained by the existence of RCMS. With 91.3 percent of people insured by RCMS people in this county have less financial barriers to health care, and the health insurance status appear to change the perception of health status.

Table 4-18 Average value of harvest and average total income per household

	Jiange	Fushuan	Yanyuan	Hanyuan	Dongpo	Total
Value of harvest	1467.19	1489.23	3882.06	2180.98	2510.89	
Total income	10318.02	14875.00	7186.03	13466.67	17054.96	11188.0

Source: ZEF survey 2002

4.3.2.2 Care-seeking decision and choice of the first providers

As indicated in Figure 4-10, there are 282 individuals in the sample reporting illness throughout last year. Of them, 89.4 percent have sought care after getting ill, whereas 10.6 percent have not sought any medical help. In terms of choice of the first contact providers, of the sick, 56.7 percent firstly visited the local public health facilities (village health station and town health center); some 24 percent visited private facilities, indicating the important role of private facilities in health care provision. The remaining 19 percent went direct to county hospital.

Care-seeking by income group

The sampled households are classified into five groups on the basis of household total income and a cross table analysis is made (see Table 4-19). The lowest, second and highest quintile are a bit more likely to seek care than the remaining groups and the difference is not statistically significant. This result demonstrates that income situation does not affect the decision of whether to seek care. However, it does affect which provider is selected. The lowest quintile of sample households is more likely to seek cheap health care (from village health stations and town health centers); the highest quintile is likely to seek care from town health center and private facilities. Exceptionally, the second lowest quintile has the highest using rate of county hospital's care. It might be interpreted that, though their diseases are only second most severe, their relatively high income may allow them to see a doctor in a county hospital. The behavior of choice of the first providers is highly correlated with the household's income ($p<0.01$).

Figure 4-10 (a) Percentage of seeking care and (b) choices of providers

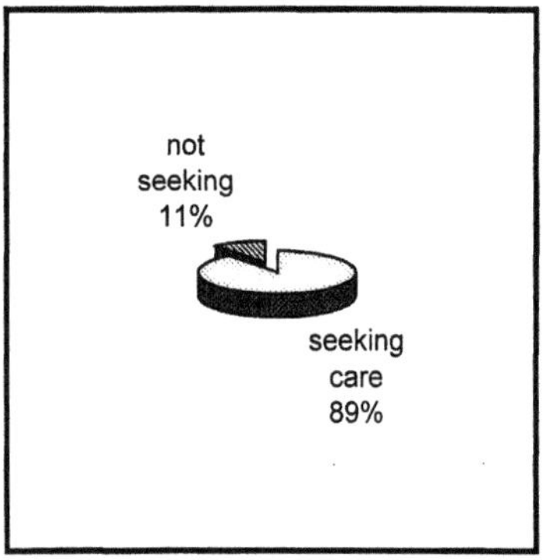

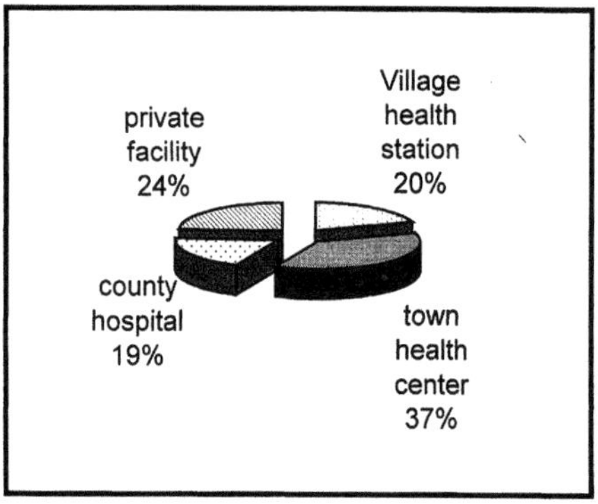

Table 4-19 Care-seeking decision, choices of providers by income quintile

Income group	Care seeking		Provider choices							
			Village health station		Town health center		County hospitals		Private facilities	
	Count	%	Count	%	Count	%	Count	%	Count	%
Lowest quintile	41	93.2	18	43.9	14	34.2	5	12.2	4	9.8
2nd quintile	40	93.0	6	15.0	16	40.0	11	27.5	7	17.5
3rd quintile	50	84.8	9	18.4	19	38.8	13	26.5	8	16.3
4th quintile	53	84.1	8	15.1	17	32.1	8	15.1	20	37.7
Highest quintile	68	93.2	9	13.0	27	39.1	11	15.9	22	31.9
Total	252	89.4	50	19.8	93	36.9	48	19.1	61	24.2

Source: ZEF survey 2002

Care seeking by counties, members of RCMS and policyholders of PHI

As stressed previously, there is a large variation in illness reports. It is expected that people in counties of low illness have a relatively high rate of seeking care reports because they appear to report relatively severe illness only. Despite the large variation of illness report among the five counties, the probability of seeking health care for those who reported illness are similar. For example, Jiange with the lowest illness reports has almost the same rate as Dongpo with the highest illness reports. The difference in the care-seeking rate between all counties is not statistically significant.

Nevertheless, there is a significant difference in selecting care providers across the five counties ($p<0.01$). An example of the rate of choosing "town health center" as the first contact provider makes this point. The rates of Fushun, Jiange, Dongpo, Hanyuan and Yanyuan are 67.6, 48.0, 40.0, 24.6 and 18.4 percent respectively. People in each of the counties have the same or similar probability of seeking care, but the providers they visit varies largely across counties. The patients in three of the sample counties, Jiange, Fushun and Dongpo, tend to seek care at town health care centers, while the patients in Yanyuan concentrate on seeking care from village health station and those in Hanyuan on seeking care in private health facilities.

As shown by the first column of Table 4-20, the percentage of seeking care of policyholders of PHI and members of RCMS is slightly higher than non-policyholders and non-members but statistically significant.The ownership of PHI policy and membership of RCMS, however, has a significant impact on selecting providers. Of policyholders of PHI, 32.6 percent sought care in county hospitals while 17.2 percent of non-policyholder sought care in county hospitals. This significant difference may mainly be caused by the regulation stipulated by insurance companies that the insured are required to see a doctor in county hospital or above except that emergency health care is allowed to be received in town health centers. As a rule, health care received in private facilities is impossible to be reimbursed. So it is not surprising that only 9.7 percent of policyholders sought care in private health facilities while 26.2 percent of non-policyholders utilize health care from here. Another finding is that the non-policyholder's choice of providers is evenly distributed among the four types of providers whereas policyholders tend to seek health care from town health centers and county hospitals.

Though only Dongpo district has RCMS, the researcher tries to make a primary examination of the impact of membership of RCMS on care seeking behaviors. The members of RCMS employ health care provided by town health centers more frequently (42.6%) than non-members (34.5%); the difference is statistically significant. Thus, the possibility of seeking care in county hospitals is largely reduced. One reason is that town health centers in towns of RCMS have been equipped with relatively advanced technologies and staffed with well-educated professionals, and hence most cases can be cured there. The other reason is that RCMS encourages its members to seek care at town health centers by repaying the medical bill at a higher rate.

Table 4-20 Care seeking and choice of providers by county, insurance status

Variables	Seeking care		Village health station		Town health center		County hospital[1]		Private facilities[2]	
	Count	%	Count	%	Count	%	Count	%	Count	%
County	p=.626		p=.001							
Jiange	25	89.3	6	24.0	12	48.0	7	28.0	0	0
Fushun	37	92.5	1	2.7	25	67.6	6	16.2	5	13.5
Yanyuan	49	92.5	24	49.0	9	18.4	11	22.5	5	10.2
Hanyuan	63	86.3	5	8.2	15	24.6	16	26.2	25	41.0
Dongpo	79	89.9	14	17.5	32	40.0	8	10.0	26	32.5
Ownership of PHI	p=.672		p=.078							
Non-policyholder	220	89.1	45	20.4	80	36.2	38	17.2	58	26.2
Policyholder	32	91.4	5	16.1	13	42.0	10	32.3	3	9.7
Membership of RCMS	p=.725		p=.021							
Nonmember	177	88.9	36	20.3	61	34.5	42	23.7	38	21.5
Member	75	90.4	14	18.7	32	42.7	6	8.0	23	30.7

Source: ZEF survey 2002

Care seeking by the severity of illness

Not surprisingly, severity of sickness measured by duration of illness affects the care-seeking behavior significantly (Table 4-21). Normally treatment mainly reduces the extent of symptoms of an acute disease (e.g., influenza) but plays a relatively small role in lowering the duration of illness. Of those who were ill below 3 days, only 81.6 per cent decide to seek care in a health facility, while of those who were ill for 30 to 90 days, 100 percent sought medical help.

Table 4-21 Care-seeking and choice of providers by severity of sickness

	Seeking care		Village health station		Town health center		County hospital[1]		Private facilities[2]	
	Count	%	Count	%	Count	%	Count	%	Count	%
Duration of illness	p=0.003		p=0.001							
0 (< 3 day)	40	81.6	4	10.0	13	32.5	3	7.5	20	50
1 (3<=d<7day)	34	87.2	5	14.7	15	44.1	4	11.7	10	29.4
2 (7<=d<15day)	58	84.1	7	12.1	26	44.8	11	19.0	14	24.1
3 (15<=d<30day)	33	94.3	9	27.3	8	24.2	7	21.2	9	27.3
4(30<=d<90day)	42	100.0	12	29.3	13	31.7	10	24.4	6	14.6
5(d>=90day)	45	93.8	12	26.7	18	40	13	28.9	2	4.4

Source: ZEF survey 2002

Note: 1. This category 'county hospital' contains other similar hospital attached to university and large- and medium-sized state-owned enterprise and so on; 2. This category is inclusive all the private health care providers: private clinic, hospital, traditional healer as well as drug store.

As far as the choice of the first providers is concerned, the longer the disease lasts, the more likely people were to go to county hospital, which was best equipped among the four types of providers. For those who were ill less than 3 days, only 7.5 percent went to county hospital, while for those who were ill above 90 days, 28.9 percent sought health care in county hospital. The severely sick seem to seek care directly in county hospital and the slightly sick seem not to seek health care or visit a private facility or village health station, where consultant fees are low relative to that of county hospital.

Care seeking behavior vs. demographic characteristics

A variety of demographic factors may affect the care seeking behavior and some are of more interest. Gender, age, education, relation to household head, ethnic group, mother's education, cigarette smoking and self rating health status are among the list of interests (see Table 4-22). Obviously, the male do not differ from the female in care-seeking decisions. But males are significantly more likely to seek care in county hospitals and private facilities compared to females. The majority of females chose town health centers while males visit the 4 types of health facilities more equally.

Table 4-22 Care seeking and choices of providers vs. individual and household attributes

	Seeking care		Village health station		Town health center		County hospital		Private facilities	
	Count	%	Count	%	Count	%	Count	%	Count	%
Gender	P=.934		p=.050							
Female	124	89.2	23	18.4	51	40.8	16	12.8	35	28
Male	128	89.5	27	21.3	42	33.1	32	25.2	26	20.5
Relation to household head	0.98		p=.033							
Household head	71	89.9	14	19.7	29	40.9	15	21.1	13	18.3
Spouse	76	90.5	9	11.8	35	46.1	11	14.5	21	27.6
Kids	77	88.5	19	25.0	16	21.1	18	23.7	23	30.3
Others	28	87.5	8	27.6	13	44.8	4	13.8	4	13.8
Education	P=.487		p=.001							
Illiterate	82	91.1	30	36.1	27	32.5	12	14.5	14	16.9
Primary school	59	85.5	8	13.8	27	46.6	8	13.8	15	25.9
Junior high school	39	92.9	6	15.4	14	35.9	7	17.9	12	30.8
Senior high school	49	87.5	5	10.2	19	38.8	12	24.5	13	26.5
College and above	20	90.9	1	5	4	20.0	9	45	6	30
Ethnic group										
Han	203	88.6	26	12.8	84	41.4	37	18.2	56	27.6
Minority	49	92.5	24	48.9	9	18.4	11	22.5	5	10.2
Mother's education	p=.258		p=.001							
Illiterate	80	87.0	31	38.8	27	33.8	13	16.3	9	11.3
Primary school	52	85.2	4	7.8	18	35.3	12	23.5	17	33.3
Junior school	49	96.1	9.0	18.0	19	38.0	10	20.0	12	24.0
Senior and above	65	90.3	5	7.7	25	38.5	13	20.0	22	33.8

Source: ZEF survey 2002

Note: % refers to percentage of the count of each cell in the total count of the row.

Education affects the choice of providers, but it does not affect care-seeking decision. The best-educated individuals are most likely to choose county hospitals (45%) and private facilities (30%) while the illiterate individuals visit village health stations and town health centers more frequently. The relation to the household head also influences the choice of providers only. The core members, especially the kids, have a greater possibility of using costly hospital services while non-core members mainly employ services provided by facilities at the levels of village and town. This result suggests the presence of discrimination against non-core members. Regarding ethnic group, minorities mainly employ services in village health care stations (49%) because they are settled less densely and have a greater geographic barrier to town health centers, not to mention by county hospitals. Mother's education affects both care seeking decision and the choice of providers. Not surprisingly, the individuals living in a home with a more educated mother are more likely to seek care and choose county hospital or private services[14].

Care seeking by distance

The distance between home and town center affects decision of care seeking and choice of the providers significantly. The patients who live farther from the town health center seem to be discouraged from seeking medical help outside. For the choice of the first providers, home's distance from center does hinder the ill from seeking care in town health centers. The closer the home is to town health centers, the more likely people are to seek care here. Some 53 percent of the closest quarter and only 21.2 percent of the furthest quarter of patients visited town health centers.

Table 4-23 Care seeking and choice of the first providers vs. distance

	Seeking care		Village health station		Town health center		County hospital		Private facilities	
Distance (4 categories)	Count	%	Count	%	Count	%	Count	%	Count	%
1(<1km)	53	85.5	7	13.2	28	52.8	10	18.9	8	15.1
2(1<=d<5km)	148	91.4	35	23.7	51	34.5	24	16.2	38	25.7
3(5<=d<9km)	18	78.3	2	11.1	7	38.9	1	5.6	8	44.4
4(d>=9km)	33	94.3	6	18.2	7	21.2	13	39.4	7	21.2

Source: ZEF survey 2002

To sum up, descriptive analysis shows that beneficiaries of private health insurances and rural cooperative medical schemes are more likely to utilize services provided by town health centers and county hospitals. This pattern can be mainly explained by the insurance effect, but it is also attributable to

[14] It is interesting to note that the education of the household head does not affect health care-seeking behavior significantly. However, the result is not presented here.

proximity of beneficiaries' homes to town centers or county seats, and hence facing lower transportation and time costs. Alternatively, the pattern may be attributed to higher incomes of the insured individuals. Therefore, without controlling income and other factors, it is not conclusive to what extent higher utilization rate of higher-quality health services is due to the insurance effect.

4.3.3 Estimates of care-seeking decision and choice of the providers

A binary logit model of care seeking decision and a multinomial logit model of choice of providers are estimated to identify the determinants in general and to detect the impact of PHI in particular. On the basis of Eq.4-31, the following 4 regressions are estimated. The first one is for BL model and the remainder for MNL model. The definitions and means of dependent and independent variables concerned are presented in Table 4-24.

Table 4-24 Definition of variables for binary logit (BL) and multinomial logit (MNL) estimation

Variables	Definition of variables	Mean BL	Mean MNL
Dependent			
SEEKCAER	Care seeking decision: 1 if one individual seeks care; 0 otherwise.	0.894	
FSTHF02	Choice of provides: 0 if individual visits private health facilities, 1 if village health station, 2 if town health center, 3 county hospital		
Independent	1) Individual attributes		
GEND	Gender: 1 if individual is male, 0 if individual is female	0.517	0.510
AGE	Age of the individual in years	36.033	35.830
EDUYEARS	Formal school education in years	4.738	4.788
CHRONDIS	Chronic disease: 1 if one has chronic disease, 0 otherwise	0.273	
ILLDURAT	Illness duration in days	61.72	63.93
	2) Household (head /spouse) attributes		
ETHNIC	Ethnic group: 1 if household head is minority, 0 if being Han people	0.196	0.203
SPOUSEDU	Formal school education of spouse of household head in years	4.439	4.535
DISTANCE	Distance from home to town center in kilometer	4.067	
	3) Access variables		
LG_INC	Logarithm of total household income in Yuan.	4.016	
LG_CSHPC	Logarithm of cash income per capita in Yuan.		7.291
LG_CHARG	Logarithm of user fees charged by health care provider in Yuan.		5.547
LG_WTIME	Opportunity costs including all treatment-related time (curing time, attentive time, travel time and waiting time)		1.888
	4) Health insurance status		
PRIHSINS	Private health insurance: 1 if one is policyholder, 0 otherwise.	0.118	0.120

$$\log\frac{P}{1-P} = \alpha + \beta_1 A_i + \beta_2 E_i + \beta_3 S_i + \beta_4 D_i + \beta_5 SE_i + \beta_6 DIS_i + \beta_7 CI_i + \beta_8 9HI_i + \mu$$

$$\log\frac{P_1}{P_0}=\alpha_{10}+\beta_{10}A_i+\gamma_{10}E_i+\delta_{10}S_i+\varphi_{10}D_i+\eta_{10}SE_i+\tau_{10}ET_i+\kappa_{10}CI_i+\lambda_{10}WT_{ij}+\theta_{10}CH_{ij}+\vartheta_{10}HI_i+\varepsilon_{10}$$

$$\log\frac{P_2}{P_0}=\alpha_{20}+\beta_{20}A_i+\gamma_{20}E_i+\delta_{20}S_i+\varphi_{20}D_i+\eta_{20}SE_i+\tau_{20}ET_i+\kappa_{20}CI_i+\lambda_{20}WT_{ij}+\theta_{20}CH_{ij}+\vartheta_{20}HI_i+\varepsilon_{20}$$

$$\log\frac{P_3}{P_0}=\alpha_{30}+\beta_{30}A_i+\gamma_{30}E_i+\delta_{30}S_i+\varphi_{30}D_i+\eta_{30}SE_i+\tau_{30}ET_i+\kappa_{30}CI_i+\lambda_{30}WT_{ij}+\theta_{30}CH_{ij}+\vartheta_{30}HI_i+\varepsilon_{30}$$

where subscript 0, 1, 2, 3 denotes private health facility, village health station, town health center, and county hospital, respectively.

4.3.3.1 Binary logit model of care-seeking decision

The parameters and corresponding standard errors are shown in Table 4-25. Most of the parameters bear expected signs. For example, the serious and rich patients and those who are insured by private health insurance are likely to seek care. However, these parameters are not statistically significant. The only significant factor affecting the decision of seeking care is mothers' education. However, the chi-square test of function shows that all the parameters are not statistically different from zero (p=0.733). This estimated results indicates that the decision of seeking care is fundamentally not affected by such factors as income and ownership of health insurance. No further analysis on the decision of whether to seek care is carried out because of the insignificance of the estimation results of whole BL model. Some explanations for this result will be available in the discussion about the impact of PHI on the choice of the first providers in next section.

Table 4-25 Estimation results of impact of PHI on care seeking: BL model

Variables	Coefficient	Standard error
GEND	0.083	0.408
AGE	-0.009	0.012
EDUYEARS	-0.030	0.060
CHRONDIS	0.542	0.529
ILLDURAT	0.003	0.002
ETHNIC	0.844	0.724
MOTHEREDU	0.114*	0.070
DISTANCE	0.009	0.062
LG_INC	0.181	0.643
PRIHSINS	0.157	0.683
Constant	0.838	2.574
Number of observations	271	
Log likelihood function	-90.842	
Chi-squared	6.921	
Significance level	0.733	

4.3.3.2 Multinomial logit model of choice of the first providers

Sequence of choice of providers

The parameters and corresponding standard errors and significance are presented in Table 4-26. The variables entering the multinomial logit model are similar to the variables in the logit model of demand for private health insurance in last section. However, the researcher uses as many continuous variables as possible instead of category variables to avoid observations disproportional to independent variables in the regression.

The results demonstrate that higher cash income makes it more likely for an individual to seek care in the sequence of private facility (PF), county hospital (CH), town health center (THC) and village health station (VHS), as seen in Table 4-27. This is basically consistent with our expectation. The individuals in a richer household tend to select the county hospital because of the higher quality (efficacy), and choose the private health facility because it takes less time to travel and to cure the disease and treatments are carried out in a friend manner. Moreover, the rich households tend to have higher wage of time and thereby their opportunity costs are higher than those of the poor. This finding is also approximate to some previous researches on choice of providers in China (e.g., Yip *et al* 1998, Henderson *et al* 1998). The individual's selecting orders with respect to significant variables are summarized in Table 4-27.

Many demographic attributes of individuals and household heads are significant in the previous descriptive analysis. But only gender is significantly related to choice of providers after controlling other factors. Being male, one is more likely to select providers in the sequence (CH > VHS >THC > PHF).

The ownership of a private health insurance plays a significant role in choosing a provider. Policyholders are more likely to seek care in the public medical system (VHS, CH and THC in turn), less likely to seek care in private facilities, other things being held constant. The coefficients of village health stations and county hospitals bear expected signs and are statistically significant. It is confusing that policyholders are a bit more likely to select village health station. Three circumstances may be related to the finding. First, village health station is the first round of selection; the patients can move to another type of health facility if it cannot meet the need for health care. Second, village health station is the cheapest and most time-saving option. Finally, the private health insurance covers only a range of relatively severe diseases and has high deductibles, thus the ill insurants may try to obtain some primary diagnostic information at low costs to see if the conditions are covered by the PHI policy. If not, they also avoid using services of county hospital that are both costly and time consuming.

The user fee is also a significant determinant of provider selection. A higher user fee makes it more likely for a person to select higher quality of services. The selecting sequence (CH>THC>VHS>PHF) of the patients in the sample households is completely consistent with the expectation.

Table 4-26 Parameter of impact of PHI on choice of providers: MNL model

Variable	Coefficient	Standard Error	P[|Z|>z]	Coefficient	Standard error	P[|Z|>z]
Village health station	(1)			(2)		
GEND	1.026**	0.479	0.032	0.938*	0.484	0.053
AGE	0.008	0.013	0.555	0.011	0.014	0.410
EDUYEARS	0.002	0.073	0.975	-0.015	0.074	0.835
ETHNIC (1=minority)	0.650	0.870	0.455	0.820	0.876	0.350
SPOUSEDU	-0.093	0.085	0.278	-0.093	0.087	0.282
ILLDURAT	0.010**	0.005	0.033	0.009*	0.005	0.040
LG_CSHPC	-1.039***	0.346	0.003	-1.052***	0.350	0.003
LG_CHARG	0.409**	0.188	0.029	0.374**	0.190	0.049
LG_WTIME	-0.379**	0.148	0.011	-0.376**	0.150	0.012
PRIHSINS				1.484*	0.878	0.091
Constant	4.794*	2.780	0.085	4.900*	2.830	0.083
Town health center						
GEND	0.369	0.378	0.329	0.294	0.382	0.442
AGE	-0.005	0.011	0.655	-0.002	0.011	0.826
EDUYEARS	0.005	0.053	0.927	-0.003	0.054	0.951
ETHNIC	-1.293	0.821	0.115	-1.170	0.824	0.156
SPOUSEDU	-0.062	0.065	0.346	-0.054	0.066	0.415
ILLDURAT	0.007	0.004	0.110	0.007	0.004	0.113
LG_CSHPC	-0.664**	0.281	0.018	-0.692**	0.289	0.017
LG_CHARG	0.447***	0.144	0.002	0.416***	0.145	0.004
LG_WTIME	-0.199*	0.114	0.082	-0.209*	0.116	0.072
PRIHSINS				0.930	0.735	0.206
Constant	3.661*	2.258	0.105	3.903*	2.330	0.094
County hospital						
GEND	1.203**	0.473	0.011	1.062**	0.482	0.028
AGE	0.006	0.014	0.672	0.009	0.014	0.530
EDUYEARS	0.117*	0.068	0.087	0.101	0.069	0.146
ETHNIC	0.874	0.900	0.332	1.020	0.904	0.259
SPOUSEDU	-0.073	0.080	0.363	-0.068	0.081	0.402
ILLDURAT	0.007	0.005	0.141	0.007	0.005	0.146
LG_CSHPC	-0.116	0.331	0.725	-0.170	0.337	0.615
LG_CHARG	0.780***	0.190	0.000	0.756***	0.192	0.000
LG_WTIME	-0.111	0.156	0.475	-0.125	0.158	0.429
PRIHSINS				1.362*	0.783	0.082
Constant	-5.010*	2.843	0.078	-4.614	2.899	0.112
No. of observations	241			241		
Log likelihood	-260.81			-261.02		
Restricted log likelihood	-325.95			-325.95		
Significance level	0.001			0.001		

Source: computed from ZEF survey 2002

Table 4-27 Selection order of providers

Variables	Sequence	Variables	Sequence
Gender (Male=1)	CH > VHS >THC > PHF	Duration of illness	VHS>CH=THC>PHF
Cash income (log)	PHF >CH >THC >VHS	User fees (log)	CH>THC>VHS >PHF
Time costs (log)	PHF >CH >THC >VHS	PRIHINS (Insured=1)	VHS>CH>THC>PHF

Source: computed from ZEF survey 2002

Marginal effect on probability of choice of providers

Until now we have explored the significance and direction of effect of each variable. For some specific variables (e.g., income) the magnitude of impact is of more concern. However, the magnitude of the impact of each variable cannot be directly seen through estimation results of MNL model because the MNL model is non-linear and hence marginal effects vary with the values of corresponding variables. The marginal effects have to be computed with a particular method. Following an approach prominently stated by Greene (2000), the marginal effects of MNL model is computed based on the following formula.

Eq. 4-36 $$\delta_j = \frac{\partial P_j}{\partial x_i} = P_j\left[\beta_j - \sum_{k=0}^{J} P_k \beta_k\right] = P_j\left[\beta_j - \bar{\beta}\right]$$

where δ denotes the coefficients of marginal effects. Normally, the marginal effects are computed at means of regressors. For convenience, it is worth noting that the same scale factor applies to all the slopes in the model. Note that for any particular x_k, $\partial P_j/\partial P_k$ need not to have the same sign as β_{jk}. The marginal effects of the four choices with respect to significant variables are computed and presented in Table 4-28. Note as well that for every row the sum of increased probabilities is always equal to the sum of decreased probabilities.

First of all, being a policyholder of PHI reduces the probability of choosing private health facilities by 17.5 percent ($p<0.10$) while enhancing the probability of choosing the three other facilities by the same scale. However, each of the three other marginal effects is insignificant. Findings suggest that the ownership of PHI only significantly elevates probabilities of choosing public facilities as a whole but does not enhance the probability of choosing each of them. This finding is partially contrary to theoretical expectation that PHI has a positive effect on care-seeking decision and encourages the rural insurants to seek high-quality care provided by county hospital. It is also inconsistent with some existing findings of the impact of community based health insurance schemes[15]. For example, Juetting (2001) estimated the impact of CBHIS on the probability of hospitalization in Senegal and found that the insured had a better access to hospital services than the non-insured. Dong *et al* (2000) found that the drug use was positively and significantly related to the membership of RCMS in China.

[15] Few empirical studies on the impact of private health insurance on the care-seeking decision are found, so only research results on CBHIS is discussed.

However, some China's empirical evidence in support of our findings can be found. For example, Yip *et al*'s MNL estimation (1998) demonstrated that both government employee health insurance schemes (GHIS) and labor health insurance scheme (LHIS) are not significantly related to the choice of the providers (village health stations, town health centers and county hospitals). RCMS, however, enhances the probability of selecting village health station largely. Henderson *et al*'s (1998) logit model estimation indicated that GHIS also has no effect on the decision of care seeking and LHIS has even a negative impact on care-seeking decision. These consistent findings have twofold implications: On the one hand, the problem of moral hazard has to some extent been precluded; on the other hand, the health insurance products or schemes in China do not significantly improve the welfare of the insurants.

The weak marginal effect of PHI policy on choice of providers in this study is attributable to the relatively high deductible and a small range of severe diseases covered. The outpatient and inpatient services due to diseases that are not stipulated in the insurance contract will not be repaid. Therefore, the ownership of PHI policy does not necessarily encourage insurants to seek high quality health care directly.

Apart from the previous reasons, another key reason why PHI does not have a significant impact on care-seeking decision is that the difference in quality between 4 types of rural health facilities is appropriately designed. Actually, the consultation fee of village health station or some traditional healers, which are incorporated in 'private facilities', is so low that most rural households can afford the consultation. The poorest of the poor might have consulted doctors in village health station, but probably cannot afford the costs of the drugs as shown in Dong *et al*'s study and costs of hospitalization services as shown in Juetting's study. In our survey, unfortunately, only the general costs of illness is investigated.

Also, some marginal effects deserve attention. In demographic attributes of individuals and household heads, being male decreases the probability of choosing private facilities and town health centers by 9.2 and 8.9 percent respectively, while it increases the probability of choosing village health stations and county hospitals by 7.9 and 10.2 percent. A one-year increase in individual's education increases the probability of choosing county hospitals by 1.6 percent while reducing the probabilities of choosing the three other providers. In terms of ethnicity of the household head, being minority may significantly reduce the probability of selecting town health centers by 44.0 percent ($p<0.05$), while raising the probabilities of the three others. As far as the severity of illness measured by days of duration is concerned, a one-day increase in duration may significantly reduces the likelihood of choosing private facilities by 0.1 percent ($p<0.10$) while raising the probability of choosing public facilities.

Expectedly, cash income has a strong effect on the choice of providers. An increase of 10 Yuan in cash income per capita raises the probability of choosing private facilities and county hospitals by 10.0 and 6.4 percent ($p<0.05$), while reducing the probabilities of choosing other types of facilities. In contrast, an increase of 10 Yuan in user fees increases the probabilities of selecting county hospitals and town health centers by 6.5 and 1.2 percent (p<0.01) whereas reducing the probabilities of choosing the other providers (p<0.01).

Another variable related to costs of health care is the total time spent on all the treatment-related activities, mainly including curing time and nursing time. An increase of one day in treatment-related time may raise the probability of choosing private health facilities and county hospitals by 3.5 and 1.1 percent, respectively. Note that the increase in probability of choosing private health facilities is over three times as large as that of choosing county hospitals. Contrary to our expectation, this finding may be linked to three basic facts. First, private health facilities include many different types: private clinics, traditional healers, drug stores, homes of private doctors; second, treatment-related time tends to be significant short in drug stores; third, the ill who choose private health facilities are more likely to be suffering from chronic diseases.

Table 4-28 Marginal effects of probability of choice

	Private facilities		Village health station		Town health center		County hospital	
Variable	Coefficient	P[|Z|>z]	Coefficient	P[|Z|>z]	Coefficient	P[|Z|>z]	Coefficient	P[|Z|>z]
Gender	-0.092*	0.095	0.079	0.153	-0.089	0.220	0.102*	0.068
Education years	-0.003	0.739	-0.005	0.569	-0.008	0.471	0.016*	0.066
Ethnicity	0.040	0.726	0.180**	0.048	-0.440***	0.002	0.219**	0.024
Mother's education	0.010	0.300	-0.007	0.485	-0.001	0.977	-0.003	0.775
Illness duration	-0.001*	0.053	0.001*	0.062	0.001	0.396	0.000	0.718
Log cash income	0.100**	0.014	-0.091**	0.015	-0.074	0.147	0.064*	0.083
Log user fees	-0.073***	0.001	-0.003	0.899	0.012	0.691	0.065***	0.003
Log time costs	0.035**	0.041	-0.034**	0.048	-0.012	0.613	0.011	0.585
Private HI status	-0.175*	0.100	0.097	0.274	0.001	0.994	0.077	0.307

Source: ZEF survey 2002

Income and price elasticity

As stated in the previous section, usually the elasticity is more desirable in measuring response to some variables, particularly income and price, since elasticity is normalized for the variables' units. Economic theory suggests that increased income causes increased purchase of most goods. Most goods have positive income elasticities and are referred to as *normal* goods. On the contrary, if a good has a negative income elasticity, it is referred to as an *inferior* good. Income elasticity also helps define whether goods are necessities or luxuries. When income elasticity is between zero and 1, this good is called a necessity and when income elasticity exceeds 1, the good is called a luxury. Health care is empirically a normal good and a necessity based on individual data other than across-country data (Phelps 1997). In this study, health care provided by four different types of facilities has a significant difference in quality. Some low-quality health care is expected to be undesirable and quantity demanded tends to fall when income rises. That is, some types may be *inferior* goods. Income elasticity can be defined as the percentage change in probability of selecting one type of health care caused by a percentage of change in cash income per capita.

The focus of health care demand studies on price elasticity is making a good assessment of policy (Hammer *et al* 1995). There are many kinds of measurements of health care quantity demanded, such as all expenditures, physician services, hospital services and nursing homes. No matter which measurement of demand is used, the elasticities are in the inelastic range, i.e, between zero to −1. In short, the consumer is not responsive to price to a large degree. Where this research is concerned, like the definition of income elasticity, price elasticity is referred to as a percentage change in the probability of selecting a certain type of provider caused by a percentage of change in user fees or time costs, respectively. The computation of elasticity is made on the basis of the following formulas (Train 2003).

$$E = \frac{f(\beta X) * \beta X}{F(\beta X)} = \frac{f(\beta X) * \beta X}{\Lambda(\beta X)} = ME * \frac{X}{\Lambda(\beta X)} = ME * X * \frac{1 + e^{\beta X}}{e^{\beta X}}$$

From Table 4-29 it can be seen that health care provided by county hospitals and private facilities is a normal good and necessity. Their income elasticities are 0.15 and 0.23, respectively, indicating that when cash income per capita of consumer rises by 1 percent, the probabilities of selecting county hospital and private facility increase by 0.15 and 0.23 percent, respectively. Correspondingly, the probabilities of selecting village health stations and town health centers decrease by 0.21 and 0.17 percent. The increased percentage is equal to the decreased percentage since people have to select one of four choices as long as they decide to seek care. The higher income elasticity of services provided by county hospital reveals that the rich can capture a larger share of benefits subsidized by private health insurance.

As far as price elasticity is concerned, the probability of choosing county hospitals and town health centers will increase by 0.15 and 0.027 percent, respectively, while that of private and village health stations decrease by 0.17 and 0.007 percent respectively if the user fees increase by one percent. Rural consumers are responsive to prices of hospital services and private services, compared to prices of the services of town health centers and village health stations. Another measurement of price, time costs, is much less inelastic than user fees.

Besides the elasticity computed on means of all observations, the researcher calculates elasticities on the basis of means of different income groups, aiming to examine whether the poor are responsive to price and income, relative to the non-poor. For simplicity, the sampled households are divided into two groups: the poor and non-poor. Results reveal that the probability of the poor choosing county hospitals decreases by 0.20 percent while that of the non-poor decreases by 0.11 percent as income falls by one percent. For the most undesirable health care (village stations) the poor are more negative income responsive. For health care provided by town health centers and private facilities, the non-poor are more income responsive, though directions of change are opposite. This analysis suggests that the poor are more income responsive for the most and least desirable health care.

In terms of price elasticity, a comparison between the poor and non-poor is complex. The poor are more price elastic with respect to services of private facility and county hospital, and are less price elastic with respect to services of village station and town health center than the non-poor. It is worth noting that the price elasticity of care of county hospital is positive and both the poor and non-poor are almost equally price elastic. This can be viewed as the additional demand for quality of health care, because health care provided by county hospitals is expected to have more efficacy.

Table 4-29 Elasticity of income, price and time costs

Variable	Private facility			Village health station			Town health center			County hospital		
	Poor	Non-poor	Overall	Poor	Non-poor	Overall	Poor	Non-poor	Overall	Poor	Non-poor	Overall
Income	0.143	0.292	0.231	-0.265	-0.169	-0.209	-0.081	-0.231	-0.170	0.203	0.107	0.148
User fees	-0.100	-0.219	-0.170	-0.034	0.007	-0.007	-0.015	0.067	0.027	0.149	0.145	0.149
Time costs	0.050	0.100	0.080	-0.097	-0.062	-0.077	0.006	-0.051	-0.027	0.041	0.013	0.024

Source: ZEF survey 2002

4.4 Summary and recommendation

This chapter explores whether the private health insurance can benefit rural people and how it can be done by investigating who is privately insured and how their care-seeking behaviors are affected by the ownership of private health insurance (PHI) policies. Results indicate that about 12 percent of people in the sample were policyholders; the largest group of beneficiaries of PHI is in school, from kindergarten kids to college students, who usually participate on a group basis. Estimation demonstrates that despite many determinants, PHI purchase is individual specific and varies largely across counties in Sichuan Province. Age, education of individual and relation to the household head are significantly related to the purchase of PHI. Among household attributes, only income and distance are statistically significant factors. In addition, risk attitude and health risk magnitude are also significant determinants. Adverse selection exists as anticipated, but it is mild. Moreover, demand for PHI is related to the effort of the supply side. Where there is a fierce competition between insurance suppliers, there is a high purchase rate.

To expand demand for PHI, the insurers should give priority to the younger, better educated and risk-averse individuals in richer households that have large portions of cash income. The physical inspection of the prospects should be strictly carried out because of the presence of adverse selection. What the governments can do is to foster the competitive insurance market and offer an abatement of taxation in some products targeted at the vulnerable groups (e.g., the children) that the governments want to subsidize.

Though 16% of the sampled individuals are members of RCMS, the participation in RCMS is only qualitatively discussed at the level of county or village. The preliminary conclusion is that the government incentives, financial subsidies, relatively densely populated villagers, middle income level, knowledge of insurance and human resources in management appear to constitute the necessary conditions for the successful implementation RCMS. Furthermore, a low-cost and effective method of collecting premium is also a prerequisite.

Health insurance affects illness reports and choice of providers but does not have an impact on care-seeking decision. However, the impact of RCMS differs from that of PHI. RCMS encourages the members to report the illness (even slight sickness) early and seek care in village health stations and town health centers. Additionally, the premium is moderate. Thus, the poor can benefit more from RCMS. PHI does not affect illness reports and care-seeking decision, but it encourages the insured to seek high-quality but costly health care provided by county hospitals. Though the premium is high compared with RCMS, the rural households can still benefit by having kids and pupils insured.

By computing income elasticity of four providers we find that health services from town health centers and village health stations are inferior goods and health services provided by county hospitals and private facilities are normal goods. Compared with the non-poor, the poor are more income responsive to the most and least desirable health care (provided by county hospitals and village health stations).

These findings suggest that although both RCMS and PHI are health care financing instruments in rural areas, RCMS is a more pro-poor tool and deserve to be translated into instant action by the communities and local governments. PHI also needs encouraging policies from the governments, for example, the tax cut for 'pupils health and accident insurance'. In order to provide adequate health care for the poor, village health stations and town health centers need to be better equipped and staffed with the assistance of RCMS funds and public investments of the governments. However, a suitable difference in quality is needed to ascertain the poor to have easy access to cheap health care.

5. Performance of the Rural Cooperative Medical Scheme and Willingness to Pay for an Improved Scheme

The term of community based health insurance scheme[16] has evolved into an umbrella term that covers a wide spectrum of health care financing instruments (Hsiao 2001; Dror and Jacquier 1999). Micro-insurance, community health funds, mutual health organizations, rural health insurance, revolving drug funds, community involvement in user fee management have all been loosely referred to as community-based health care financing mechanisms. As stated in chapter 3, there are internationally four often-encountered forms of community-based health care financing mechanisms (Jakab *et al* 2001): community involvement in user fee collection, community prepayment schemes or mutual health organizations, provider based community health financing and community driven prepayment schemes attached to social insurance or government-run schemes. Community-based health insurance scheme has been recognized as a promising institutional arrangement and has been attracting widespread attention for its potential role in providing improved health insurance schemes and easy access to health care for the population segments that government and market-based health financing arrangements failed to reach (e.g., the rural, the urban poor).

China's rural cooperative medical scheme (RCMS) actually belongs to community prepayment scheme. Beginning in 1955, China was the first large nation in the world to develop a nationwide CBHIS to fund and organize preventive, primary care and secondary health services for its rural population (Liu *et al* 1996). It was rapidly developed in the 1960s and 1970s; at the pride there were over 90% of the rural population being insured. However, it is on decline since the economic reform in the 1970s. The enrolment rate of RCMS has been below 10% since the early 1980s (Figure 5-1). Although governments and communities have been eager for raising the enrolment rate of RCMS, rural population has a common concern over the performance of RCMS. So there is an imperative to reassess the performance of RCMS in achieving the goal of well and equal health (WHO 2000). Moreover, the past performance may have a direct impact on the willingness of households to join and to pay for the scheme.

Performance refers to the extent to which the rural cooperative medical system achieves its goals. The goals of RCMS can be set at multiple levels. At individual scheme level, the performance is referred to as capacity of resource mobilization and financial protection; at the level of administrative unit (township, county or nation) it is referred to as the improvement of health status.

---

[16] However, it is also called community-based health care financing mechanisms in some literatures. So community based health insurance schemes and health care financing mechanisms are used interchangeably in the section

This chapter has two major sections. In the first place, the performance of RCMS is assessed through several approaches. In the second, households' willingness to join and to pay for a RCMS with hypothetically better performance is investigated and estimated.

Figure 5-1 Time series of enrollment rate of RCMS and per capita income

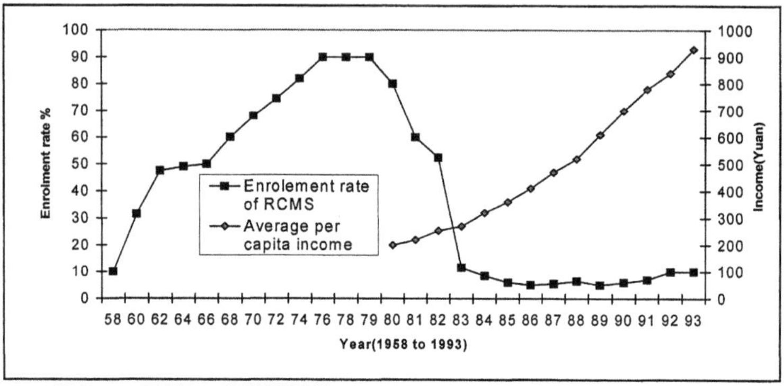

Source: Kahn *et al* 1997 and MOH 2001

5.1 Performance of the rural cooperative medical scheme

5.1.1 An overview of performance of RCMS in China

There has been no consensus on the overall enrolment rate of rural cooperative medical scheme in China. Qiu *et al* (1999) offered a wide range of percentage from 10 to 15 percent. Khan *et al* (1997) reported that only some 6 percent of the rural inhabitants were members of RCMS in 1990. Yu (2002) believed that the share of RCMS members in the total population might be as low as 4 percent. The officially publicized proportion of RCMS members was 6.6 percent of the total population and the proportion of villages with RCMS was 17 percent in 1998 (MOH 1999). According to a National Survey on Rural Health System (NSRHS) conducted by the China Academy of Social Science (CASS) in 2001, the proportion of rural households covered by RCMS had reached 11.6 percent and the proportion of villages 13.7 percent by 2000. The main reason for the lack of accurate statistics is that RCMS is not treated as a formal insurance organization by governments and need not be registered in authorities concerned.

The proportion of population covered by RCMS is quite low; an even more serious issue is that the limited RCMS is not uniformly distributed across different types of rural areas. As shown in Table 3-7, the first type of rural areas that is most developed in China has an average enrolment rate of 22.2 percent, while in the fourth type of rural areas that is least developed only 1.8 percent of people are RCMS members.

The main intention of establishing RCMS is to mobilize additional resources for health care and provide a financial protection for the members. If we want to assess the performance of RCMS, the most important two aspects of the performance should be the capacity of resource mobilization and effectiveness of financial protection against costs of illness (Arhin-Tenkorang 2001). He also innovatively advanced a systematic approach to appraising the performance of an individual community health care financing scheme. In terms of resource mobilization capacity, the main indicators includes (1) premium; (2) costs of health care consumed by the members; (3) magnitude of external subsidy; (4) size of risk pool; and (5) degree of economies to scale that is achieved. Alternative indicators measuring capacity of resource mobilization of schemes are (i) positive and greater than unity ratio of "health care expenditures (HCEXP)" to "revenues from contribution (REV)", and (ii) significantly greater than unity ratio of "average expenditure on an individual (AVEXP)" to "average individual contribution (PREM)". That is to say:

(i) Ratio 1 = HCEXP / REV > 1

(ii) Ratio 2= AVEXP / PREM >>1

In terms of risk protection effectiveness, the following indicators are supposed to be investigated: (1) wealth/asset profile of members (and non-members); (2) utilization rates of members and non-members in whole covered area; (3) rate of cost recovery; (4) size of risk pool. Arhin-Tenkorang (2001) also suggested that the risk protection effectiveness of a scheme was a function of income of members, price of health care and other goods, illness incidence, and completeness of an insurance benefits package. However, information on all of these indicators may not be readily acquired, so the following alternative indicators are proposed: Affordability of premiums; appropriateness of payment schedule; nature of co-payment; completeness of benefits package; increase (or any observable change) in utilization rate of health services due to the scheme.

5.1.1.1 Resource mobilization

China had achieved a high level of public health in the 1960s and 1970s owing to successful preventive programs and an emphasis on primary health care and universal accessibility (Bogg *et al* 1996). Rural cooperative medical scheme also contributed significantly to the resource mobilization available for local health care system (Dakab 2001) and hence largely promoted the construction of the three-tier health care delivery network. Many studies reported the contribution of RCMS to the operational revenues of town health centers and village health stations, irrespective of its premium level. A few examples are summarized in Table 5-1 by using the Arhin-Tenkorang approach. An immediate consequence of the collapse of RCMS was the destruction of village health stations, exercising a far-reaching influence on health status of rural residents, particularly for the rural poor people in remote and marginal areas. For village health stations were usually built in the center of each village and to some extent subsidized by the collective fund of each village, rendering an easy geographical and financial access to the villagers. Conversely, the destruction of village health station set up great physical and financial barriers to health care for rural population. The rural might turn to town health center, but households that are far from town center are affected disproportionately. The fact that the households living in remote and marginal areas are usually the poorest makes the bad situation worse. Additionally, of the survived village health stations, about 50% have been privatized (MOH 1999). Although the households do not need to travel long to see doctor, the sharp rise in price may still exclude a fraction of patients from the village clinics and the preventive care is usually disregarded by the for-profit private clinics. The giant role of RCMS in supporting operation of village health stations can be explicated by a case study in a state-nominated poverty-stricken county (Zhang 1997). Even though the premium is as low as 2 Yuan per capita per annum and the benefits package includes only 7 percent of cost recovery as well as exemption from fees of consulting, injection and diagnosis, the town based RCMS helped set up three, recover seven and improve 28 village health stations. The establishment of the grass root health care delivery network laid a foundation for effective and efficient rural health care provision.

Nevertheless, the capacity of resource mobilization varies tremendously across schemes in different regions (eastern, middle and western regions). RCMS in eastern-region provinces such as Shanghai and Jiangsu have mobilized relatively large amount of resources, while schemes in the middle and western region provinces such as Hubei, Sichuan and Guizhou are not capable of mobilizing that much (Table 5-1). For example, the average per capita contribution in Dongpo district of Sichuan province (6.3 Yuan) is less than 10 percent of that of Shanghai (87.5 Yuan). The external subsidies from villages and local governments in Dongpo take up 23.6 percent of the total premium income while the external resources in Shanghai account for 39.7 percent. Moreover, the small amount of government subsidies in middle and western regions often cannot reach the insurance fund account in time owing to financial deficit or other problems[17].

5.1.1.2 Financial protection

High health expense is a major cause of poverty in rural areas (Liu *et al* 1996). Literature review indicates that rural cooperative medical schemes reduce the out-of-pocket payments by 20-60% and the disease-induced poverty by 20-70% (e.g., Tan 1997; Yin *et al* 1998; Hong 2002). It also increases health care utilization of its members to a certain extent. The visiting rate and hospital admission rate enhances by 30-148% and 15%, respectively. A few excellent

[17] The data on Shanghai derivates from Gao Ge *et al*, Survey and Analysis of Funding and Compensation of the Rural Health Care Scheme in Shanghai. *China's Rural Health Administration*. Vol.22, 2002 (2). In Chinese.

studies on RCMS are summarized as follows.

Chen *et al* (2000) endeavored to determine the level of disease-induced poverty by calculating average medical expenditures for those diseases which were classified as contributing to a high financial burden based on income level and disease type. They found that disease-induced poverty decreased from 23% to 4% in Shouguang county, and from 30% to 3% in Pingdu county, Shandong province after the introduction of RCMS coverage. Carrin *et al* (1999) assessed "ratios of insurance protection" in China's rural cooperative medical system. They measured the ratio of average health insurance contribution per capita to health care expenditure per capita. Wide variation was found in the level of insurance protection across counties from as low as 10% in Lingwu and Xiaoshan to as high as more than 30 percent in Yihuang county.

Table 5-1 Performance indicators of RCMS between eastern and middle-western regions of China

	Eastern region		Middle region	Western region	
	Shanghai (1999)	Wujiang, Jiangsu (1997)	Zaoyang, Hubei (1998)	Dongpo, Sichuan (1999)	Qinglong, Guizhou (1999)
Resource mobilization					
Premium per capita	87.5 Yuan	26 Yuan	15 Yuan	5 Yuan	2 Yuan
%of premium in income	2.0%	1.0%	1.0%	0.2%	0.5%
External subsidies (Yuan)	29.4, village 5.3, government	5, village 5, government	0	1.5, government	3, Poverty reduction fund of the World
% of subsidies in total premium	39.7%	36.6%	0	23.6%	60.0%
Size of risk pool (persons)	253123	500048	219333	343491	14063
Insurance fund (10 000 Yuan per county)	2214.83	2160.36	329 per township	17174.54	38706 Yuan
Financial protection					
Affordability	Affordable to majority	Affordable to majority	Affordable to majority	Affordable to majority	Affordable to minority
Enrolment rate	70%	70%	49%	60%	10.2%
Completeness of package	OP and IP	OP and IP	IP only	OP and IP	OP and IP
Cost recovery rate	OP 20-25%; IP 30-60%;Ceiling 10000-20000 Y	n.a.	IP, 50-70%, ceiling 1000 Y	30-60%, ceiling 800 Y	30%, ceiling 200 Y
Level of access	Increased: insured/non-insured charge=2.48	Increased: after/before visits=1.36	Increased: visit rate by 36%; admission rate by 15%	n.a.	n.a.
Assessment of financial risk protection (FRP)	Significant FRP	Significant FRP	Significant FRP	Significant FRP	Insignificant FRP

Source: computed from Gao *et al* 2002; Zheng *et al* 1998; Tan 1997; Jiang *et al* 2003; Lai 2001.

Note: OP and IP are the abbreviation for outpatient and inpatient services, respectively. "N.a" means "not available"

A few researchers compared the health care utilization of RCMS members with that of non-members in China. Gu *et al* (1995) found that farmers covered by RCMS used more medical care with the same costs (3.7 times outpatient visits per farmer per year), compared with those who pay for medical care out of pocket (3.1 times). Liu *et al* (1996) showed that the cost per visit of the uninsured (3 Yuan per visit) is twice as large as that of the insured (1.5 Yuan per visit). They also found that the higher average charge per outpatient visit for the uninsured can be attributed to the fact that "these schemes can exercise their bargaining power in demanding discounted prices or the providers can be paid on a partial capitation basis" (Liu *et al* 1996; Hsiao 2001).

Another study based on household data compared out-of-pocket expenses and utilization of those who were members of the pilot RCMS with those who were not (Gu *et al* 2000). The average fee per outpatient visit was 10.1 Yuan for RCMS members, compared to 21.7 Yuan for non-members. In the mean time, higher utilization of medical care among RCMS members was observed in two pilot sites. Hospital admission rate was 60 percent among members, compared to 43 percent in the control group.

A 300-household survey aimed at appraising the sustainability of RCMS and the relation between RCMS and alleviation of farmer's economic burden was conducted in Yangzhou, Jiangsu province (Zhou 1999). He found that out-of-pocket payment of RCMS members is lower than that of non-members. Also two types of RCMS, i.e., welfare-risk-type and risk-type, were compared. On average, the out-of-pocket payments of welfare-risk-type and risk-type RCMS were 62.6 and 119.9 Yuan per capita per year, respectively, while that of non-member was 140.0 Yuan. The payments of welfare-risk-type RCMS members, risk-type RCMS members and non-members represented 1.7%, 3.4% and 3.5% of net annual income per capita, respectively. In addition, Zhou (1999) compared the average values of prescriptions of patients who were RCMS members with those of patients who were not; the average values are 3.2, 4.3 for members under welfare-risk-type, risk-type RCMS, respectively, while the average value of description for non-members is 24.9 Yuan.

5.1.2 Assessment of the Performance of RCMS in China

Three methods are employed to analyze the performance of rural cooperative medical schemes in China: with-without comparative analysis between counties (horizontal comparison), before–after comparative analysis of the country as a whole (vertical comparison) and analysis of individual schemes.

5.1.2.1 With-without comparative analysis between counties

The researcher uses the dataset of the National Survey on Rural Health System (NSRHS) at county level to do the comparative analysis. There are 30 sampled counties in the dataset, of which there are 13 counties with enrollment rate of 40% or above (for simplicity, this type of county is named RCMS county, otherwise non-RCMS county). We are concerned about the indicators of health outcome, health infrastructure, fiscal expenditures and income in the course of assessing the performance of RCMS. These 4 types of indicators in RCMS counties and non-RCMS counties are summarized in Table 5-2. Prior to the analysis of the role of RCMS in resource mobilization and risk protection there is a need to outline the basic relationship between the 4 types of indicators. Income in terms of GDP per capita is the rooted resources, which can be used to invest in health, but a high GDP does not sufficiently or necessarily guarantee high health outcomes; the net income per farmer per annum is a more accurate indicator to measure the resources disposable to health care of the rural residents. The fiscal expenditure indicators reflect the latest distribution of fiscal revenue to the health sector. The health infrastructure indicator is crucial to judging to what extent the county government spent on health improvement in the past few years. Meanwhile, the health infrastructure (particularly at the level of township) is positively associated with the establishment of RCMS. With a stable stream of consumers as the result of the rules of RCMS, these town health centers could earn more and be equipped better than it would be otherwise. The health outcome is the function of various health inputs purchased with income. Due to the lack of data on medical care consumed at the county level, the researcher is to circumvent health inputs and resort to average net income per farmer, the public expenditure on health and health infrastructure. Table 5-2 shows that RCMS counties have much higher health outcome than the non-RCMS counties. The infant mortality rate (IMR) and maternal mortality rate (MMR) of RCMS counties is 67 percent, 20 percent of that of the non-RCMS counties, respectively. Also, the counties with over 40 percent of population insured have higher life expectancies (LE). Paired sample t-test shows that IMR, MMR and LE of RCMS counties are significantly different from those of non-RCMS counties. Although significant difference in health status between RCMS counties and non-RCMS counties can be detected, this difference need not result from the high percentage of RCMS members. Only after the effect of other factors are controlled can the effect of RCMS be definitely identified. Thus, a regression analysis is required. The effect of RCMS on health outcome can be identified in the health production function by introducing the variable of RCMS. The variable of RCMS can be defined in categorized or continuous pattern. In terms of categorization, RCMS counties, where over 40 percent of the rural have been enrolled as member of RCMS[18], and non-RCMS counties are used as a dummy variable.

$$lgH_i = a_0 + a_1lg(NI_i) + a_2lg(WFEXP)_i + a_3lg(MCEXP)_i + a_4lg(RURDR)_i + a_5lg\,(DRNK)_i + a_6\,(DCMS)i + \varepsilon_i \quad \text{(Eq. 5-1)}$$

where H_i = health outcome (IMR, MMR) in county i;

[18] Because the highest enrolling rate of RCMS at the county level is 80 percent, 40 percent is defined as the cutoff line.

NI_i = average annual net income per farmer in county i;
WFEXP = welfare expenditure per capita in county i;
MCEXP = public expenditure on health (mainly including fund for family planning, public medical care, preventive medical care, women and child's health care and investment in improving rural drinking water);
RURDR = number of licensed rural doctors per 1000 person in county i;
DRNK = ratio of people accessibale to clean drinking water in county i;
DCMS = dummy variable of RCMS (1 if 40% or above of rural population is insured).

Table 5-2 Comparison of indicators of RCMS and non-RCMS counties

Indicators	RCMS county Mean	Non-RCMS county Mean	Ratio*	Significance
Income				
GDP per capita (Yuan)	9497.02	3681.51	2.58	0.001
Net income, yearly average per farmer (Y)	3356.21	2105.55	1.59	0.001
Fiscal expenditures				
Subsidy to RCMS per capita** (Y)	2.29	0.81	2.83	0.247
Social welfare expenses per capita (Y)	14.73	7.51	1.96	0.004
Per capita fiscal spending on health sectorY	12.67	6.72	1.89	0.028
Health infrastructure				
Hospital beds per 1000 persons	1.72	1.48	1.16	0.320
Health professional per 1000 persons	26.47	23.36	1.13	0.396
Licensed rural doctors per capita	0.51	0.75	0.67	0.283
Ratio of people of clean drink water	0.89	0.65	1.36	0.046
Number of village health stations	477.46	383.94	1.24	0.332
Health outcome				
Infant mortality rate (per 1000 live births)	13.14	20.68	0.64	0.015
Maternal mortality rate (per 10000)	2.73	12.46	0.22	0.081
Average life expectancy (years)	72.08	69.72	1.03	0.003

Source: computed from NSRHS by CASS 2001
Note: *Ratio=indicators of RCMS counties/ indicators of non-RCMS counties.
**Subsidy to RCMS per capita is computed by dividing aggregates subsidies to RCMS by the total population rather than the number of members in the county.

Estimation results shown in Table 5-3 demonstrate that over 40 percent of the rural insured by RCMS does not make a significant difference in health outcome after the control of effect of income, whether in terms of IMR or MMR. Moreover, the health outcome is closely related to the farmer's net income, and government's public expenditure on health and welfare.

Because all the dependent and independent variables (except a dummy) are logarithm, the coefficients can be interpreted as elasticities. Elasticities of annual net income per farmer to IMR and MMR are –0.8 and –1.99, respectively, indicating that a one-percent increase in annual net income per farmer may lead to a decrease of 0.8 percent in IMR and a decrease of 2 percent in MMR. The welfare expenditures per capita contribute significantly to the decrease in IMR. A one-percent increase in welfare expenditures lowers IMR by 0.24 percent. It is noteworthy that the public spending on health care per capita bears a sign contrary to the expectation that an increase in public spending on health should lead to a decrease in MMR. A possible explanation is that many counties made a remedial fiscal spending in the survey year because of the lower IMR and MMR in the past years. The increased public spending in health cannot enhance the IMR and MMR immediately. Thus what we observed is this contradictory phenomenon based on the cross sectional data.

Table 5-3 Estimation results of health outcome indicators: log-linear model

Dependent variables	IMR		MMR	
	Coefficients	t	Coefficients	t
Constant	4.06***	4.97	7.12**	3.42
Log welfare expenditures per capita	-0.24**	-2.63	-0.20	-0.85
Log average annual net income per farmer	-0.80***	-3.29	-1.99**	-3.20
Log public spending on health care per capita	0.05	0.80	0.34**	1.85
Log licensed rural doctor per capita	-0.12	-1.60	-0.37	-1.09
Log ratio of people with clean drinking water	0.10	0.74	0.32	1.00
Dummy of 40% villagers of RCMS	0.01	0.10	-0.05	-0.27

Source: computed from NSRHS by CASS 2001

To check the robustness, the ratio of members of RCMS is measured continuously and the log-linear model is re-estimated. The result presented in the Appendix B exhibits that the percentage of RCMS members in different counties does not contribute significantly to health status improvement at the county level in 2000.

In conclusion, the ratio of RCMS members to the local population does not sufficiently explain the improvement of IMR and MMR as the effects of other factors are controlled. It is also important to note that welfare expenditure is as important as net income per farmer in lowering IMR and that public spending on health care is as important as net income per farmer in reducing MMR.

5.1.2.2 After-before comparative analysis of the whole country

Many studies indicated the possible close relation between the breakdown of the rural cooperative medical system and the deterioration of rural health care conditions in China. For example, Hsiao (1995) found that China's health status improvement between 1952 and 1982 resulted partly from RCMS including the three-tier health care delivery network. The World Bank (1997) clearly pointed out that the under-five child mortality rate (U5CMR) in China declined steadily until the early 1980s and then stagnated until 1991. The decline and stagnation of U5CMR coincided with the breakdown of RCMS. Experiences from other countries suggested that U5CMR did not need to plateau as China's in the late 1980s (as measured by census and survey data).

A 300-poverty-stricken-county study seems to confirm this relation (Liu *et al* 1996). Despite the low economic growth rate, the poverty-stricken counties had experienced steady economic growth since the economic reform. Per capita

GDP in real term increased in the late 1980s. However, the health status of people in poverty-stricken counties showed a sign of deterioration. For example, median IMR in the surveyed poor counties increased from about 50 to 72 per 1000 live births during the same period of time.

A more important indicator investigated by the National Statistics Bureau (NSB) consecutively is IMR. As shown in Table 5-4, the change in IMR ratio of the rural to urban population does not coincide with the change in income ratio of the rural to urban. Though the gap in income between the rural and urban was narrowing, the gap in IMR broadened from 1.65 in 1981 to 3.35 in 1991. This phenomenon emerged after RCMS broke down in the period from 1979 to 1981. Obviously, the private resources in households cannot fully explain the enhancement of IMR. A new explanatory factor—public resources mobilized through RCMS is then introduced.

To investigate the role of RCMS in enhancing the health status of rural people, one way is to compare the scale of external resources mobilized through RCMS before its collapse with that afterwards. However, there is not the item of resources mobilized in the Statistics Yearbook. This comparison has to be made indirectly. Though RCMS broke down in 1981, there is likely to be a time lag between the collapse and the change of the health indicators and hence it is rational to use indicators a few years after the collapse. The availability of data is another consideration. Thus, health indicators and incomes in 1981 with those in 1991 are compared.

Table 5-4 Infant mortality rate (IMR) in China's urban and rural areas

Year	Rural area (deaths per 1000)	Urban area (deaths per 1000)	IMR ratio of rural to urban	Per capita income ratio of rural to urban
1981	40.5	24.5	1.65	1:2.57
1991	58.0	17.3	3.35	1:2.40
1995	41.5	14.0	2.96	1:2.50
2000	37.0	11.8	3.14	1:2.79

Source: NSB (2002); MOH (2002)

Before the comparison, some fundamental assumptions should be made. Health production is theoretically the function of medical care and consumption of other goods. Empirically, income that represents the constraint of the health production function is widely used as an explanatory variable. Many studies have shown a positive correlation between income and health status, whether on a cross sectional or longitudinal basis (Jack 1999 and literature cited therein). In addition to farmer's own resources, the governments can also mobilize some external resources for health inputs through particular fiscal expenditure, public program and RCMS. Thus, the possible private health resources measured by income per capita and public resources measured by resources mobilized from external sectors are two main explanatory variables of health outcome. The health production function is assumed to be a log-linear function of income per capita and average money per capita mobilized from other sectors (see Eq. 5-2).

It is also implicitly assumed that technology coefficient of health production holds constant within the time period such that all the difference in health outcome is created by the difference in health inputs, whether over time or across urban and rural areas.

Eq. 5-2 $$\lg H = c + \alpha \lg I + \beta \lg S$$

where H = health indicator, here it refers to IMR in particular
I = income per capita
S = resource mobilized from external sectors
c = technology coefficient (e.g., education and knowledge of mother)
α and β are the parameters of income and external money respectively.

Based on the features of production function, α, β should be more than 0 and less than 1. To verify the larger resources per capita (S_b) mobilized from the external sector before 1981 than after the breakdown (S_a), it is needed to transform the equation (5-2), using H and I to express S. The proof of larger external resource mobilized can be seen in Appendix B. Results of calculation show that the urban/rural ratio of public resources mobilized through health care financing mechanism in 1991 is higher than that in 1981. There are two possibilities for the higher ratio in 1991: much larger resources were mobilized for the urban health sector or much less resources mobilized for the rural health sector in 1991. It is known that health care financing mechanisms in urban areas had almost not changed from 1981 to 1991, but health care financing mechanisms in rural areas had changed dramatically. Thus, the conclusion can be drawn now that there is a larger share of public resources in total health resources mobilized into the rural health sector before the breakdown of RCMS than that afterwards. RCMS did play a significant role in narrowing the IMR gap between the urban and rural before the economic reforms in the early 1980s.

5.1.2.3 Case study on the performance of RCMS in Sichuan

In reality, the constraint that the coefficient of health production function holds constant is not necessarily tenable; then it is relaxed in this section. The insignificant role of rural cooperative medical scheme in achieving health improvement on the county level should not deny the role of individual RCMS in resources mobilization and financial protection. A low efficiency of health production may easily waste the additional resources mobilized. Therefore, a study on the performance of an individual RCMS, offering a more accurate assessment, is needed.

A few RCMS surviving the economic reforms make the individual scheme study possible. The researcher conducted a focus interview with two officials who were in charge of the administration of RCMS in Meishan prefecture and Dongpo District (originally named Meishan county). Dongpo district is a medium-sized, medium-developed county located on the Chengdu Plain. By the year of 2001, the total population was 8.15 million and the rural residents accounted for 73.7% of the total; GDP per capita was 5279 Yuan ($638.33) and average annual net income per farmer was 2524 Yuan. As seen in Table 4-1, the health infrastructure in this county is good, relative to other sampled counties. There were 33 towns and 56 town health centers. On average, each town had nearly two health centers. The hospital beds and health professionals per thousand people were 1.8 and 2.6, respectively, which were very high, compared with other counties in Sichuan province. Moreover, every village has functioning health station. RCMS under persisting operation may underlie the relatively good health infrastructure and intact three-tier health care network.

Meishan County has been famous for its health insurance experiment undertaken by Rand Corporation of the United States in collaboration with Ministry of Health of China in the 1980s, whereby it became the model of implementing RCMS. One of the interviewees took part in that experiment and has been engaged in the administration of RCMS since then. The major interview results can be summarized as follows: first of all, the numbers of towns/townships with RCMS and of RCMS members were on the decline. There were 19 towns in 1999 and 14 towns in 2000 with functioning RCMS in the Dongpo district. The number of RCMS members diminished from 399 thousand in 1999 to 343 thousand in 2000 as shown in Table 5-5. Secondly, the ratio of the premium of RCMS to annual net income per farmer is lower than 1%, compared with that of the social health insurance of urban employees (6%). Thirdly, the enrollment rate was quite high (96%) as long as a town had RCMS. Fourthly, the benefits package involved health inspections for members in those households without hospitalized members over the past two years, enabling the direct benefits for all households participating in the schemes. Finally, 7.6 million Yuan was mobilized through RCMS; 6.8 million Yuan was indemnified to 11890 inpatients in 1999 (2.9%). However, the total premium income was 2.15 million Yuan in 2000.

Following the Arhin-Tenkorang approach mentioned previously, the researcher calculated some indicators on performance of individual schemes presented in Table 5-7. Results indicate that either the capacity of resource mobilization or effectiveness of risk protection is weak. In terms of capacity of resource mobilization, for example, RCMS in the town of Shiqiao charged the members premium of 5 Yuan per capita per year; the county and township governments subsidize 0.5 and 1 Yuan per year per member, respectively; the risk pool in Shiqiao RCMS includes 10,378 members. The costs of health care consumed by members and degree of economies to scale were unavailable. However, a small portion of premium failed to be collected for unknown reasons; subsidies of town government also failed to be paid in the past two years because of the financial crisis in rural areas.[19] Two alternative indicators

[19] There was a Rural Cooperative Fund Union in each of the towns in Dongpo district, which was initially founded for supporting the village and township enterprises by the county Agricultural Administration and was free from the strict regulation of the Central Bank. Thus,

measuring the capacity of resource mobilization of schemes were also computed. The ratio of "health care expenditure" to "revenue from contribution" is 1.04 and the ratio of "average expenditures on individual" to "average individual contribution" should be as high as 1.25. In reality, the ratio was only about 1.10 in the town of Shiqiao in 2000. The small difference between the ratio and the unity indicates that the capacity of resource mobilization is quite weak.

Table 5-5 RCMS in Dongpo District of Meishan Prefecture, 2000

Name of towns	Premium (Y/year. capita)	Subsidy (Y/year. capita)	Size of risk pool	Enrol-ment rate (%)	Theoretical total premium	Practical total premium	Ratio of PTP to TTP	Practical contributi on per capita
Xianger	6	1.0	13950	96	97650	78120	0.80	5.60
Shangyi	4	1.0	32362	93	161810	126210	0.78	3.90
Duoyue	5	1.0	27861	96.5	167166	100299	0.60	3.60
Baimapu	5	1.0	20772	97	124632	105937	0.85	5.10
Jinfeng	8	2.0	9711	98	97110	85456	0.88	8.80
Chongre	5	1.0	15141	95	90846	78127	0.86	5.16
Yuexing	5	1.0	32491	96	194946	169603	0.87	5.22
Taihe	5	1.0	34266	98	205596	174756	0.85	5.10
Fuxing	4	1.0	21680	93	108400	84552	0.78	3.90
Luoping	5	1.0	17326	96.5	103956	83164	0.80	4.80
Xiuwen	5	1.0	29586	96	177516	133137	0.75	4.50
Wanshen	5	1.0	25947	98	155682	132329	0.85	5.10
Longxin	5	1.0	9983	97	59898	47918	0.80	4.80
Shiqiao	5	1.0	10378	98	62268	55000	0.88	5.30
Sum	6.25	1.1	343491	96	2146818	1717454	0.80	5.00

Source: ZEF survey 2002.
Note: TTP, PTP represent the theoretical total premium and practical total premium, respectively.

In terms of financial protection, RCMS is rather ineffectual. Firstly, the premium is quite low relative to annual net income per farmer; in most townships the ratio of premium to net income per farmer is only 0.2%, much lower than the rate (1-2%) recommended by the WHO, not to mention the rate of urban employee health insurance scheme (6%). Furthermore, the low premium can afford only low coverage, which subsequently brings about a low willingness to pay for this type of scheme. Secondly, the rate of cost recovery was 55 percent in Shiqiao town. As shown in Table 5-6, 505 members in total sought inpatients services and were eligible for claims in 2000. The total sum of reimbursement was 56,922.4 Yuan. The average inpatient reimbursement was 112.7 Yuan, and the average inpatient costs was 204.9 Yuan that is calculated on the basis of the sub-sample of Sichuan in the NSRHS. The highest repayment in this town

the problem of large amount of non-performing loans arose. For fear of a financial crisis in the context of the Asian Financial Crisis in 1997, the State Council announced that all the Fund Union had to be closed down.

amounted to 1800 Yuan last year. Thirdly, the benefits package of RCMS basically resembled that of a private health insurance policy. The costs of AIDS, and injures caused by attempts to commit suicide and crime, for instance, could not be repaid while the costs of inpatients due to childbearing were covered, but the rate of cost recovery declined by 20 percent. The low enrolment rate and coverage was attributable to lack of law to guarantee this scheme. Moreover, there has even been conflict between regulations from the Ministry of Agriculture and the Ministry of Health since 1999. The Ministry of Agriculture issued a regulation to ease the financial burden of farmers, which prohibits the communities and local governments from compulsively raising fees from farmers except those that were listed on the farmer's burden card.

Table 5-6 Reimbursement and inpatients in the town of Shiqiao, 2000

Village	No of inpatients	Amount of reimbursement	Average reimbursement
Xuguan	217	23807.0	109.71
Putian	82	9573.0	116.74
Tiandian	85	9152.4	107.68
Yushan	66	7448.0	112.85
Gaodeng	55	6942.0	126.22
Sum	505	56922.4	112.72

Source: ZEF survey 2002

Table 5-7 Performance of health insurance scheme in the town of Shiqiao

Resource mobilization capacity indicators	Value	Effectiveness of risk protection indicators	Value
Premium (Y/Year. capita)	5	Ratio of premium to net income per farmer	0.2%
Costs of health care consumed by the members (Inpatient)	113866.8	Payment schedule	Once a year, after the harvest
Magnitude of external subsidy	0.3Y per member	Rate of cost recovery	55%
Size of risk pool	10378 persons	Size of risk pool	10378
Ratio 1=HCEXP/REV	1.04	Completeness of benefits package	Like private health insurance
Ratio 2=AVEXP/PREM	1.10	Co-payment rate	45%

Source: computed from ZEF survey 2002

5.2 Willingness to pay for the improved RCMS in Sichuan province

In chapter 4 the observed demand for the private health insurance policy has been estimated and analyzed in detail, and also the observed demand for rural cooperative medical scheme has been qualitatively analyzed. However, owing to lack of ideal data applicable to econometric analysis, factors that affect demand for RCMS and the extent to which they affect the demand remain unknown until now. Moreover, in the process of RCMS extension, the health administrators and village cadres are always puzzled over what type of RCMS would be viable and preferable. To address these issues the contingent valuation (CV) approach is applied in this section.

The contingent valuation approach was firstly utilized by Davis in his study on environmental economic issues in 1964. But the number of applications did not grow rapidly until the middle 1970s (Kriström 1990). The approach had then been extended to other disciplines such as health economics, educational economics. Contingent valuation is a survey-based method frequently used for placing monetary values on environmental goods and services not bought and sold in the market (Carson 2000). The essence of CV approach is to ask people to provide in a hypothetical setting a monetary value for a good for which the market is missing. The advantage of CV method is that it can create hypothetical markets for non-market goods and public services so that the non-market goods' prices are readily observed and hence make them possible for economists to conduct cost-benefit analyses and for agents to make sound decisions. In the mean time, some economists argued that real transactions rendered much more reliable indicators of value than self-reported behavioral intention (e.g., Randall 1997). Suspicion about the reliability arose because of "hypothetical answers to hypothetical questions" which may deviate from the "true" valuation to a certain extent. Put it concretely, the deviation comes mainly from 'response effect biases' (Carson 1989), which encompass three groups: (i) incentives to misrepresent response, (ii) implied value cues and (iii) scenario misspecification. The first group of response effect biases comprises strategic bias and compliance bias. The second type involves the elements of the created market experiment that conveys a 'true' value of the good to the respondent and influences the response. The final group of response effects includes a number of effects related to the scenario as put forward by the researcher and perceived by the respondents. For example, the researcher specified a scenario that is hard for the respondents to imagine and then the respondents just muddle with the questions.

As far as the validity of the CV method is concerned, many careful examinations of validity have been done. For example, Bishop and Heberlein (1979) compared actions in hypothetical hunting-permit market (selling/buying hunting permits) with a real market (the permits were traded using real payment), the estimated consumer surplus per permit in real market was lower than the willingness to pay in the hypothetical market. These estimates can also be sensitive to the choice of estimation techniques; Loomis (1990) and Hanemann *et al* (1989) compared the validity of the CV method across time, their results provided additional support for the notion of consistency of valuation across experiments. After an overall theoretical and empirical examination, Carson *et al* (2000) concluded that many of the alleged problems with CV could be resolved by careful study design and implementation. The Nobel Laureates Kenneth Arrow and Robert Solow *et al* (1993) in their panel report submitted to the National Oceanic and Atmospheric Administration (NOAA) general counsel concluded that "CV studies can produce estimates reliable sufficiently to be the starting point for a judicial or administrative determination of natural resource damages—including passive use values". The researcher argues that the CV method is a very useful instrument to assess the market value of non-market goods or services.

The contingent valuation method will be employed in this section to estimate the potential demand for RCMS in China and identify the factors affecting the willingness to pay (WTP). This approach has already been widely used in the area of health economic studies. The studies included WTP for disease treatment and management, new technology, and efficacy evaluation of health care and health programs (e.g., Walraven 1996; Zethraeus 1998 and Liu *et al* 2000). However, there are few studies dealing with the issues of health care financing in developing countries. For instance, Asenso-Okyere *et al* (1997) used the bidding game method to assess the willingness of households in the informal sector to join and pay premiums for a proposed National Health Insurance Scheme in Ghana. About 64% of the respondents were willing to pay a premium of $3.03 monthly for households with five members. Asfaw (2002) also estimated the willingness to pay for a proposed community based health insurance scheme in seven rural districts in Ethiopia by the means of bidding game. Results exhibited that some 95 percent of the respondents were willing to join the scheme and the highest willingness to pay was 1.42 Euro per month per household (Birr 10). Another study for rural health insurance was conducted in India (Mathiyazhagan 1998). The heads of 1000 households in rural areas were asked directly how much they were willing to pay for a community based health insurance scheme. In the study, WTP for different types of medical benefits was estimated. It was found that most households selected a comprehensive medical care benefit and were willing to pay Rs. 163.48 per year.

Theory and empirical evidences suggest that studies undertaken in developing countries obtain valid and reliable health-related WTP results (Russel *et al* 1995). The most commonly used methods have been the bidding game, the payment card and the take-it-or-leave-it (TIOLI) (Diener *et al* 1998; Klose 1999). Nevertheless, careful design and procedural implementation are essential to increase the validity and reliability to a satisfactory extent.

5.2.1 Conceptual and analytical framework

The initial objective of CV method is to estimates values for public goods that are typically sub-optimally allocated by free operating markets. For example, the biodiversity provided can benefit anyone in the area, enjoyment by one cannot detract from enjoyment by others. Hence, someone may free ride. As a consequence, the available funds will be insufficient for optimal provision. Why CV is used for health insurance demand valuation? Although community-based health insurance scheme is not a pure public good, the market is missing in rural areas. *Revealed preference* method in private insurance market can provide accurate price of a product that is similar to the CBHIS on the basis of morbidity and costs of illness episodes (actuarially fair price). The shortcoming of this way is that it takes account only of direct use (covering costs of illness) while ignoring the indirect use (e.g., the psychological security). So the welfare gain of a health insurance scheme tends to be under estimated. Another weakness is that the actuarial calculation does not produce any information on demand. So this premium (actuarially fair price plus loading fee) is scientific but not necessarily optimal. *Stated preference* method (here referring to CV), which can theoretically measure the total economic value, will provide not only a valuation of the total value of CBHIS but also useful information for the forthcoming execution of the hypothetical health insurance scheme, e.g., whether more than half of the villagers agree with the implementation and whether they are able or willing to pay a certain amount of premium.

Theoretically, WTP is at least not less than the actuarially fair price of the insurance scheme as long as the respondents, like an actuary, offer the WTP on the basis of the costs of illness discounted by probability of disease of their households. However, it is in practice possible that the WTP is lower than actuarially fair price. Under this situation, subsidies of the government should be available if the government intends to achieve the health equity between the rural and the urban.

5.2.1.1 Theoretical model of willingness to pay

Willingness to pay (WTP) can be defined as the maximum amount that individuals are willing to pay in order to acquire a certain benefits package provided by the hypothetical health insurance scheme. This sum is the amount of money that would make an individual indifferent between contributing to the scheme and subsequently getting reimbursement when health care costs incurred and not contributing to the scheme while keeping the money. There are basically two paradigms for the interpretations of CV method response: the random utility paradigm (Hanemann 1989) and the expenditure function paradigm (Cameron and James 1987; Cameron 1988). However, McConnell (1997) show that the two methods are often similar in practice. So the researcher prefers to use the first. Household utility is defined as a function of the consumption of a compound good *Z* and the security of health insurance scheme, *HI*, given by:

Eq.5-3 $$U = U(Z, HI) \qquad Uz > 0 \text{ and } Uhi > 0$$

In this case the utility is increasing with *Z* and *HI*. Solving the consumer problem, we can obtain the indirect utility of an individual who joins the community based health insurance scheme given by:

Eq.5-4 $$V_1 = V_1(Y - WTP | HI > 0, X, \mu)$$

where *Y* is household income, *WTP* is a money amount that makes *HI* non zero, *X* is a vector of the household's observable characteristics and μ is a vector of non-observable characteristics. In technical terms, *WTP* refers to the amount of money that equates indirect utility functions with and without the existence of a certain scheme. Mathematically,

Eq.5-5 $$V_1(Y - WTP^*|HI > 0, X, \mu) = V_0(Y|HI = 0, X, \mu)$$

Next, assuming that $V_1(\lambda, WTP|HI > 0, X, \mu)$ is increasing in λ, there exists an inverse function $\psi(v, X, \mu)$ which makes $\psi(V_1(\lambda, WTP|HI > 0, X, \mu), X, \mu) = \lambda$. Therefore, the willingness to pay can be expressed as:

Eq.5-6 $$WTP^* = Y - \psi(V_0(Y|HI = 0, X, \mu), X, \mu) \equiv WTP^*(Y, X, \mu)$$

5.2.1.2 Decision procedure and econometric specification

The rural households' decision to pay for a hypothetical RCMS could be decomposed into two steps. First, households have to make up their mind to join or not on the basis of the expected utility of two options, namely, willingness to join (WTJ); next, conditional on the decision to join they decide how much to pay for a specific benefits package, namely, willingness to pay (WTP). Based on this two-step process, the WTJ and WTP's decision models are going to be estimated independently. The researcher assumes that $v\,(.)$ is linear in the parameters but log-linear in household income and μ is a logistic symmetric error distribution with zero mean. Thus, a logit model is applied to simulate the willingness to join (WTJ).

Eq.5-7 $$\log \frac{\Pr ob(join)}{1 - \Pr ob(join)} = \alpha_i \log(Y_i) + \beta_i X_i + \mu$$

where X is the vector of household attributes, and Y is household income.

A multiple regression equation is used to model how much one is willing to pay for the hypothetical RCMS once one decides to participate in the scheme, namely, $WTP > 0$.

Eq.5-8 $$(WTP)_i = \alpha_i Y_i + \beta_i X_i + u_i$$

The explanatory variables in equation Eq.5-8 are not necessarily just the same as those in equation Eq.5-7.

Alternatively, a less frequently used method is to derive WTP values from estimated health care demand equations. However, unless unconditional health care demand functions are estimated this method suffers from adverse selection problems since demand functions are basically estimated on the basis of the data from those who already had some health problems and sought health care.

5.2.2 Data

5.2.2.1 The design and administration of WTP survey

There are a variety of WTP questions applicable to the research. Open-ended, bidding game, payment cards, binary response and binary response with follow-up are among them. Different forms of questions have different merits and demerits. However, an open-ended questions format is designed in this study to ask the respondents directly how much they were willing to pay for the benefits package provided by the proposed CBHIS. The merits of the open-ended format are that (i) the answer can be directly taken as the maximum WTP; (ii) WTP can be measured as a continuous variable and hence is easy to be analyzed with simple statistical methods such as OLS; (iii) it is efficient and hence requires a relatively small sample. Its demerits encompass the difficulty in answering, high non-response rate, unreliable answers and high liability to strategic response behavior of respondents (Asfaw 2002). The reason behind the demerits is a frequent claim that familiarity with a good is a prerequisite to providing "meaningful" responses to CV valuation questions. The rationale given for needing familiarity is the assertion that respondents cannot have well-defined preferences in economic sense for goods, of which they have no direct experience. However, no standard microeconomic text has ever stated that prior experience is a precondition to rational decision-making (Carson *et al* 2000). The first three weaknesses will to a large extent be overcome in the study, because the proposed health insurance scheme is fully familiar to respondents. RCMS, as known to all, ran well in rural China in 1960s and 1970s. In two sample towns there still exists the similar health insurance scheme. Compared with the existing scheme, the proposed scheme promised a lower co-payment (half and zero) and broader coverage of diseases and injuries, which make it more desirable for respondents. The strategic bias due to free-rider can also be negligible because the health insurance scheme is not a pure local public good, i.e., it can exclude those who do not pay from enjoying the benefit brought by the CBHI scheme. The fourth weakness should be avoidable with well-trained, neutral interviewers (Carson *et al* 1994). Furthermore, Green, McFadden and Kanheman (1998) show that even open-ended questions could be incentive compatible, while a large number of economists tend to rely more on the binary approach (Carson, Grove and Machina 1999; Arrow *et al* 1993).

As argued by some economists (Carson *et al* 2000), many of the alleged problems with CV can be resolved by careful study design and implementation. Much care is taken of designing the CV questions and implementing the survey. According to a user guide of CV (Carson 2000), the wording of the CV question modified from the work of Asfaw (2002) is depicted as follows:

"As you know nobody is free from the risk of illness or injury. It is also clear that when somebody is ill/injured her/his ability to generate income will decrease and consequently s/he will not be able to pay for medical treatment and to buy medicines. Consequently, both the patient and her/his family will be in a very great trouble. Therefore, it is quite logical to help one another at the time of illness rather than crying and spending a lot of money on funeral services. In urban areas all the employees have been insured by urban employee basic health insurance scheme.

Assume that a community based health insurance scheme will be established in your village. A community based health insurance scheme is something like the former cooperative health medical scheme, which will be established voluntarily by the contribution of its members and helps its members to get free medical care when they are ill or injured. The government (Ministry of Health) will help organize the scheme. The members themselves will monitor the operation of the scheme."

The insurance scheme will provide the following services if your household joins it: (1) when you or your household members are ill or injured the scheme will cover both outpatient and inpatient costs of treatments and medicines at the government and NGO health facilities. (2) However, the insurance will not cover costs if treatment is sought from private health care providers.

In the wording of CV questions a vague depiction of the proposed scheme is avoided; second, a health insurance scheme is not a simple good, it is essential not only to describe its role in coping with financial risks of illness and benefits one can enjoy, but also to make the operative mechanism known to respondents. To help respondents to have a quick and clear scenario of the scheme, RCMS, which is quite similar to the proposed scheme and familiar to respondents, is mentioned as a reference. Not only we are aware of the accurate wording in creating the hypothetical good and its market but also we take notice that questions should be put in a suitable place within the entire questionnaire. It is placed on section 7 after the cluster information, household member's social-economic characteristics, extended family's gift exchange, health status and health care seeking as well as health risk coping with health insurance are inquired. The previous inquiries may render the respondents able to perceive the function of the hypothetical scheme naturally and associate the value with their health risk probabilities, magnitude of health care costs, substitute goods and past experience and hence figure out their own willingness to pay with less effort and more ground.

Next, we translated the questionnaire into Chinese by using very common words and phrases but consistent with terms in health care and health economics. Interviewers were then recruited from senior college students specializing in agricultural economics or industrial-commercial management. Apart from their excellent comprehensive performance in study and character, another requirement is that they come from the sample county so that each is able to speak the local dialect, which will facilitate the communication with the respondents. For interviewers to receive training is a very important link of the chain. The training contents include the general knowledge on the research problems, objectives, significance of the research as well as how to implement the inquiries. The interviewers were required to strictly comply with the original wording of CV question, not adding and subtracting any words from it to avert bias from implied value cues. If the respondents have not followed a question after the first statement of the question, the interviewers should repeat it from beginning to end time after time until the respondents fully catch the points. The researcher together with interviewers also carried out a pilot survey in six households: in the first place, the researcher demonstrated an interview; in the second place, each interviewer went through a survey independently; and finally we summarized the lessons in the survey and exchanged their experiences with each other in a meeting. These measures guaranteed that the similar scenario of health insurance scheme, if not the same, would be presented to respondents and avoid the large regional variation in WTP due to interviewer bias. The respondents were also given adequate time to consider the hypothetical good. To be more informative, the respondents who are unwilling to join or still in hesitation were asked to provide some reasons.

5.2.2.2 Sample

This study is based on a Health Insurance Survey in Rural S ichuan (HISRS) conducted at the beginning of 2002. The study areas of WTP include five sampled counties in Sichuan Province. The researcher sampled 300 households out of 10 villages randomly. The villages, towns and counties were sampled in the manner stated in Chapter 4. Despite three choices of answers, 'join', 'not join' and ' I am in hesitation', with respect to WTJ, the third choice is treated as the equivalent to the second one for convenience of the dichotomous analysis, partly because it has the same nature as the second one; partly because there are too few who responded enumerators with the third answer.

5.2.3 Results and discussion

5.2.3.1 Willingness to join and its determinants

Percentage of WTJ and reasons for unwillingness to join

As shown in panel (a) of Figure 5-2, the results demonstrate that, of 300 households, 208 (69.3%) are willing to join the hypothetical RCMS immediately, while 16.0 percent are unwilling to join (UWTJ) and 14.7 percent still in hesitation (IH). The most frequent reason cited by respondents is 'distrust of the management of the insurance fund', which accounts for 36.7 percent of the reluctant (UWTJ & IH). The second frequently cited reason is non-affordability, 25 respondents (27.8%) belong to this group. The remaining r easons are "no disease" (21.7%), "self management" (8.9%) and "family help" (4.4%), as shown in panel (b) of Figure 5-2. The percentage of WTJ is not optimistic as expected considering that many of the sampled villages each had once established a successful RCMS. Also, it is much lower than percentage of WTJ in Ethiopia (94.7%), while slightly higher than that of Ghana, 64 percent (Asfaw 2002; Asenso-Okyere *et al* 1997).

Selection and description of factors affecting the willingness to join

The researcher already made a qualitative analysis of factors that may affect the realistic demand for rural cooperative medical scheme in chapter 4. But an econometric analysis of demand for RCMS was not undertaken because all the members of RCMS are distributed in Dongpo county. Now the randomly distributed data allows a quantitative analysis of determinants of demand for a hypothetical improved RCMS. The household head's attributes such as sex, age, ethnic group, main occupation and education largely affect the expected utility function. Considering the traditional division of labor that household heads (they are assumed males) pay their main attention to money earning and housewives think more about consumption and risk management, housewives' characteristics (e.g., education) could be important determinants. Despite the fact that a great proportion of income comes from non-farm labor and sideline business, the land area should still be a factor to predict the WTJ, especially when the time of premium payment is designed after the harvest.

Figure 5-2 Percentage of WTJ (a) and reasons for unwillingness to join (b)

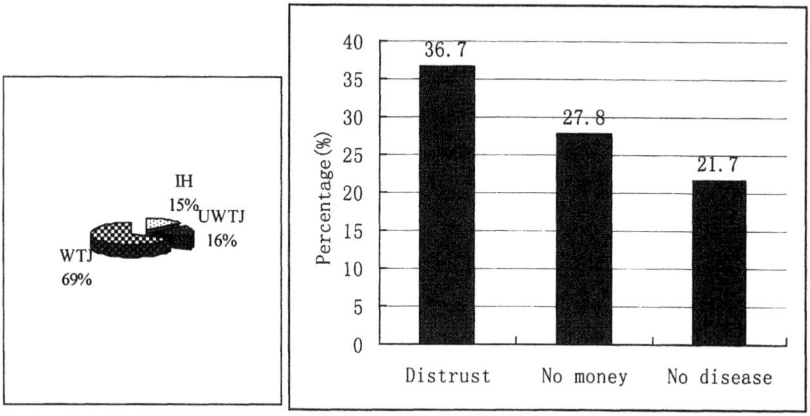

(a) (b)

The classical theory of health insurance suggests that adverse selection is inevitable because insurance policy buyers have an informational advantage over sellers. Insurance buyers know better than sellers about their particular health status or expected demand for health care. When applying this to RCMS, rational household heads on behalf of households may predict the expected demand for care on the basis of historical record of the amount of health care costs. Thus, high-risk households in terms of total health care costs in the past years are usually more inclined to join the scheme than low-risk ones.

It was primarily the patronage, corruption and poor management of insurance funds that contributed to the downfall of former RCMS (Liu *et al* 1996). These lessons may produce a profound negative influence on willingness to join in the hypothetical insurance plan. Some researchers also found rural people's distrust in the insurance fund's management in other developing countries (Wiesmann *et al* 2000 and literatures cited therein). A testable implication of the distrust is whether the experience of RCMS is helpful to improve willingness to join.

Before the estimation, the frequency of discrete variables and their correlation to WTJ are explored by means of *contingency table*. The results are presented in Table 5-8. Most of the households' attributes except occupation of the household head are significantly related to WTJ. One remarkable point is that education of housewife positively and significantly affects WTJ, whereas household head's education does not. The finding confirms the common hypothesis that women's education contributes more to a household in terms of development. Unexpectedly, the past experience of RCMS largely lowers willingness to join the hypothetical improved RCMS. The category of respondents who said they remembered RCMS clearly are less likely to join than the categories of "who do not know RCMS at all" and "who only heard of it". This result suggests that some problems with the past RCMS existed. Whether RCMS is a substitute of private health insurance policy concerns many policy-makers and researchers. Results do not provide any evidence in favor of that. There appears to be a bit conflict between private health insurance purchase and willingness to join the hypothetical RCMS, but it is not statistically significant.

Table 5-8 Cross Table of Willingness to join the improved RCMS vs. factors

Variable		N	%	Significance
Head sex	Female	5	41.7	0.005
	Male	203	70.5	
Occupation of head	Farmer	156	70.6	0.251
	Non-farmer	52	65.8	
Ethnicity	Han	156	65.0	0.004
	Minority	52	86.7	
Head education	Illiterate	41	71.9	0.400
	Primary education	65	77.4	
	Junior high school	50	64.1	
	Senior high school and above	52	65.0	
Mother's education	Illiterate	71	69.6	0.081
	Primary education	54	81.8	
	Junior high school	35	61.4	
	Senior high school and above	46	68.7	
Private HI	Purchased	55	64.0	0.309
	No	153	71.5	
Wealth status	Poor	40	63.5	0.001
	Average	102	67.5	
	Rich	65	80.2	
Migrant worker	With	82	61.7	0.042
	Without	124	75.2	
Experience of RCMS	Once with RCMS	94	65.7	0.002
	No	113	72.9	
Knowledge of RCMS	Not at all	97	70.3	0.002
	I heard of it	82	75.9	
	I remember clearly	29	54.7	
County	Jiange	51	85.0	0.000
	Fushun	24	40.0	
	Yanyuan	52	86.7	
	Hanyuan	34	56.7	
	Dongpo	47	78.3	

Source: computed from ZEF survey 2002

Ethnic group that household heads belong to is worthwhile to be emphasized because the Yi nationality lives in their own community and has their own culture. The Yi is one of 55 minorities and resides in a separate prefecture with high autonomy. Yi people are characterized by honesty, cohesion, trust and selflessness though they live in poverty. They are thought to have more social capital within the community according to the definition of social capital by Coleman (1994). Yi villages have fewer functions of state and higher degree of autonomy because of the autonomous policy for minority regions. Accordingly, they resemble communities in western countries more than Han villages. Through this variable the effect of social capita could be somewhat tested.

Income is usually of much interest. Income when measured by self-judgment compared with fellow villagers has a very significant impact on WTJ. The 'richer' group is much more likely to join than 'average' and 'poorer' groups.

Econometric analysis of factors

All the variables that are used in the econometric analysis are defined in Table 5-9. A logit model is estimated through three specifications and the estimation results are presented in Table 5-10. In the first model county difference and risk aversion are not introduced. Estimation results demonstrate that seven variables are identified as significant factors affecting willingness to join. Five of the seven variables bear expected signs: ethnicity, household head's education, mother's education, land area, cash income and whether a household has migrant workers. For instance, the Yi people are much more likely to be willing to join the hypothetical RCMS. The more households earn cash income, the more likely they are willing to join. However, age and education of the household head bear signs that are inconsistent with theoretical expectation that the aged and educated are generally more likely to join the hypothesized scheme. The possible reason for bearing opposite signs may be that the aged and educated heads are more capable of making use of alternative risk management instruments such as private transfers within the extended family[20].

As stated before, we cannot use the coefficients of the logit model to make a marginal analysis because the equation is nonlinear. The coefficients of marginal effects are computed on the basis of the equation (4-18). The marginal effects of significant variables in logit models are also shown in Table 5-10. Now the probability of willingness to join as a result of a small change in any independent variable can be directly read. For instance, a one-year increase around the mean of mother's education would lead to an increase in probability of WTJ by 2.8 percent. Likewise, as for dummy variable "ethnicity", being minority would enhance the probability of WTJ by 34.8 percent. With respect to cash income, the marginal effect is actually the elasticity of WTJ because both

[20] Or they simply rely on the financial support of their offspring to settle the medical bill.

the dependent and independent variables are in the form of logarithm that consists with the standard definition of elasticity. It can be explained that a one-percent increase in cash income will lead to 0.2 percent of increase in WTJ. As the comparison of all the parameters shows, income is a very strong determinant indicating that the rich are more likely to join.

Table 5-9 Definition of variables for WTJ Binary Logit model

Variables	Definition of variables	Mean
Dependent var.		
WTJ_01	Willingness to join: 1 if one is willing to join; 0 otherwise	0.694
Independent var.		
Head attributes		
ETHNICT	Ethnic group dummy: 1 if the household head belongs to minority; 0 otherwise	0.214
HEADAGE	Age of household head in years	42.923
HEADEDU	Education of household head in schooling years	5.439
SPOUSEDU	Education of house wife in schooling years	4.203
Household attributes		
HHSIZE	Household size: number of members of a household	4.148
INDRADE	Ratio of independents to dependents	2.689
JOBBER	Jobber dummy: 1 if a household has a member being migrant worker; 0 otherwise	0.458
DISTANS	Distance of home from the town center in Kilometers	4.040
Income		
LANDAREA	Land area owned by a household in Mu	7.217
LG_INCCS	Logarithm of cash income of the household last year (Yuan)	3.748
Health risk magnitude		
LG_HLCST	Logarithm of health care costs of the household last year (Yuan)	1.705
Experience		
CBHISEXP	Dummy: whether a household has any member having once joined RCMS	1.528
Risk aversion	Reference group=non smoking	
CIGAR_FW	Dummy: 1 if household head smokes below 1 packet a day and 0 otherwise	0.638
CIGAR_MN	Dummy: 1 if household head smokes over 1 packet a day and 0 otherwise	0.192
County	Reference group=Dongpo county	
CTY_JG	County dummy: 1 if one household is in Jiange and 0 otherwise	0.177
CTY_FS	County dummy: 1 if one household is in Fushun and 0 otherwise	0.192
CTY_YY	County dummy: 1 if one household is in Yanyuan and 0 otherwise	0.214
CTY_HY	County dummy: 1 if one household is in Hanyuan and 0 otherwise	0.203

After the reform at the beginning of the 1980s, a huge number of rural labor forces that were surplus corresponding to the rural land swarmed into the costal cities to do casual jobs. Several tides of migration from the rural to urban areas have emerged since the economic reforms. Of 300 households in the sample, 133 reported at least one migrant worker now or in the near past (44.6%). Estimation results indicate that the migrant jobber is significantly negative related to the willingness to join. The evident reason is the inconvenience that these migrant workers working in other places reimburse the medical bill. More importantly, the high health care costs occurred in coastal cities make the small amount of repayment less attractive.

Table 5-10 Estimation results of WTJ the hypothetical RCMS: Logit model

Variable	Coefficient	Std. Error	Marg. Coef.	Coefficient	Std. Error	Marg. Coef.	Coefficient	Std. Error	Marg. Coef.
	(1)			(2)			(3)		
Head attributes									
ETHNICT	2.286***	0.708	0.348	2.470***	0.746	0.370			
HEADAGE	-0.028*	0.017	-0.004	-0.025	0.017	-0.004	-0.034**	0.018	-0.006
HEADEDU	-0.155***	0.059	-0.024	-0.148**	0.059	-0.022	-0.195***	0.066	-0.034
SPOUSEDU	0.183**	0.056	0.028	0.187***	0.056	0.028	0.193***	0.062	0.033
Household									
HHSIZE	0.042	0.162	0.006	0.026	0.163	0.004	0.158	0.170	0.027
INDRADE	0.063	0.057	0.010	0.063	0.057	0.009	0.062	0.061	0.011
JOBBER	-0.933***	0.323	-0.142	-0.951***	0.326	-0.142	-0.822**	0.359	-0.143
DISTANS	-0.049	0.041	-0.008	-0.050	0.041	-0.007	-0.029	0.046	-0.005
Income									
LANDAREA	0.209***	0.057	0.032	0.216***	0.059	0.032	0.034	0.050	0.006
LG_INCCS	1.250**	0.532	0.190	1.292**	0.536	0.194	1.781***	0.590	0.309
Health risk									
LG_HLCST	-0.039	0.108	-0.006	-0.040	0.108	-0.006	-0.103	0.119	-0.018
CBHISEXP	-0.295	0.346	-0.045	-0.287	0.347	-0.043	-0.938*	0.510	-0.163
Risk aversion									
CIGAR_FW				0.343	0.413	0.051	0.489	0.466	0.085
CIGAR_MN				0.091	0.531	0.014	-0.486	0.574	-0.084
County									
CTY_JG							1.270	0.780	0.220
CTY_FS							-2.202***	0.544	-0.382
CTY_YY							3.095***	0.991	0.537
CTY_HY							-0.994	0.654	-0.172
Constant	-3.127	2.149	-0.476	-3.723	2.267	-0.558	-3.179	2.535	-0.551
Number of observations	271			271			271		
Log likelihood function	-135.090			-134.683			-120.213		
Restricted log likelihood	-165.289			-165.289			-165.289		
Chi-squared	60.398			61.212			90.153		

Source: computed from ZEF survey 2002

Note: 'Std. error' and 'Marg. coef.' denote standard error and marginal coefficient, respectively.

Next, let's turn to specification (2). The risk aversion measured by the numbers of cigarettes the household heads smoke everyday is introduced into the model. All the variables still bear the same signs and have the similar significance. The risk aversion measured by cigarettes smoking does not impact the willingness to join significantly. The reason may be that the measurement is not accurate enough to reflect the difference of risk attitude of a household. In addition, the introduction of new variables can be viewed as a test of sensitivity. As a result, the model is stable because all the coefficients almost do not change after the introduction of new variables.

Furthermore, it is believed that the difference of county may partially explain the variation in probability, thus the dummies of county are introduced into the model as seen in the specification (3). Estimation results demonstrate that the introduction of county dummies increases the log likelihood function and value of Chi-squared largely. It is worthwhile to note that the variable of ethnicity is excluded because of the high collinearity with county variables. Dongpo county being the reference group, households in Jiange and Yanyuan counties show significantly large WTJ, while households in Fushun county show significantly small WTJ the hypothetical RCMS.

5.2.3.2 Willingness to pay and its determinants

The mean and distribution of WTP

To see what types of benefits package are preferred and demanded most, three packages are designed, namely, inpatients services coverage with no co-payment, complete coverage with half co-payments and complete coverage without co-payments. As shown in Figure 5-3, the sampled households are willing to pay an average of 19.57 Yuan per capita per year for "coverage of inpatient services without co-payment". For "complete benefits package without co-payment" the sampled households are willing to pay on average 58.08 Yuan per capita per year and for "complete benefits package with half co-payment" they are willing to pay 32.50 Yuan per capita per year. One noteworthy point is that WTP for a "complete benefits package with half co-payment" exceeds half the WTP for a "complete benefits package with no co-payment". This indicates that the RCMS with complete benefits package and half co-payment is more viable and acceptable in the setting of rural Sichuan.

The distribution of WTP is also very important besides average value. Figure 5-4 indicates that for a hypothetical RCMS with half co-payment the distribution of WTP is skewed left and concentrates on the range from 1 to 25 Yuan per capita per year, which accounts for more than 50% of those who are willing to join and pay. For a scheme with no co-payment the amount of WTP lies mainly in a range of 25 to 50 Yuan, which accounts for nearly 50%; meanwhile, more than 20% of those who express the willingness to join would like to pay above 100 Yuan. Despite the difference of WTP for different benefits packages (co-insurance rate), the tendencies of change as income rises are almost the same. However, the figures of tendency are not shown in the thesis.

Figure 5-3 Mean of willingness to pay for different benefits packages

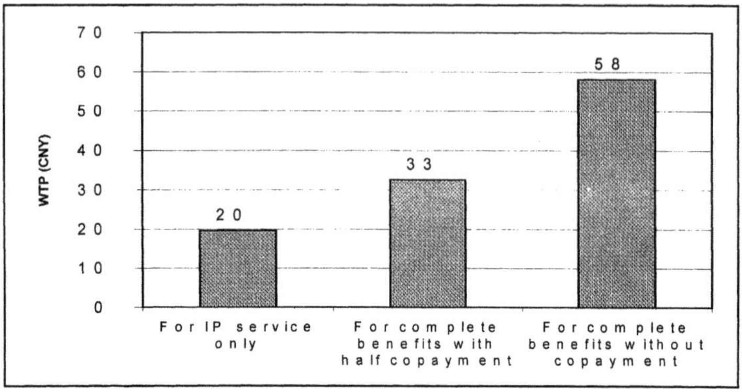

Figure 5-4 Distribution of WTP for complete benefits with half and with no co-payment

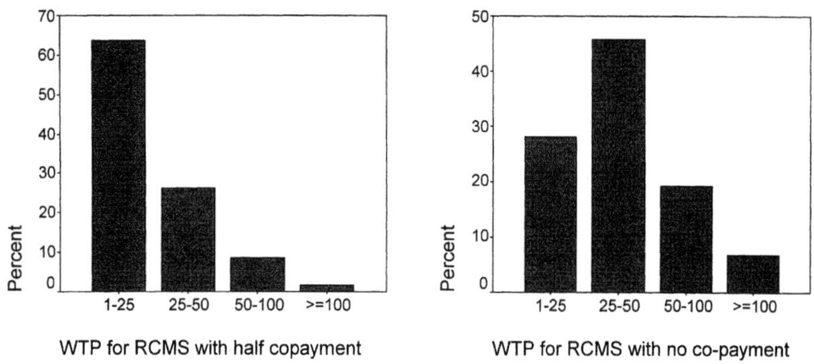

Variation of WTP with income by different categorization of other variables

So far determinants of WTP for different benefits packages and to what extent these variables determine WTP have been identified. The researcher is of further interests in how WTP varies with the rise of income at a certain level of another variable (e.g., mother's education). This question is equivalent to decomposing the joint effect of income and mother's education. To fulfill this task, it is needed to do some transformation of the two variables as follows. First of all, continuous variables are categorized. Namely, household income is divided into 10 deciles and the mean WTP in each income group is calculated; education is classified into illiterate, primary education, junior high school, and senior high school and above. Then the variation of WTP for 'complete benefits package with *half* co-payment' with income is plotted at each level of mother's education (Figure 5-5). It is generally shown that the mean WTP goes up in a spiral pattern as income rises from the lowest deciles to the highest at any level of mother's education. WTP of the groups of "elementary school" and "junior high school" does not vary considerably with the increase of income. On the other hand, at the same income level, it is discovered that households with better-educated mother are willing to pay more, except that the group of some primary education is willing to pay ever less than the illiterate group as income are relatively high.

Concerning the mean WTP for a complete benefits package with *no* co-payment, a similar tendency of WTP is observed. As the former, the WTP spires as income rises at any level of mother's education. The curves of the latter even have the same number of wave crests and troughs as the corresponding curves of WTP for a complete package (Figure D-1 of Appendix D).

Figure 5-5 Variation of WTP with income at different mother's education

- **WTP for a complete package with *half* co-payment**

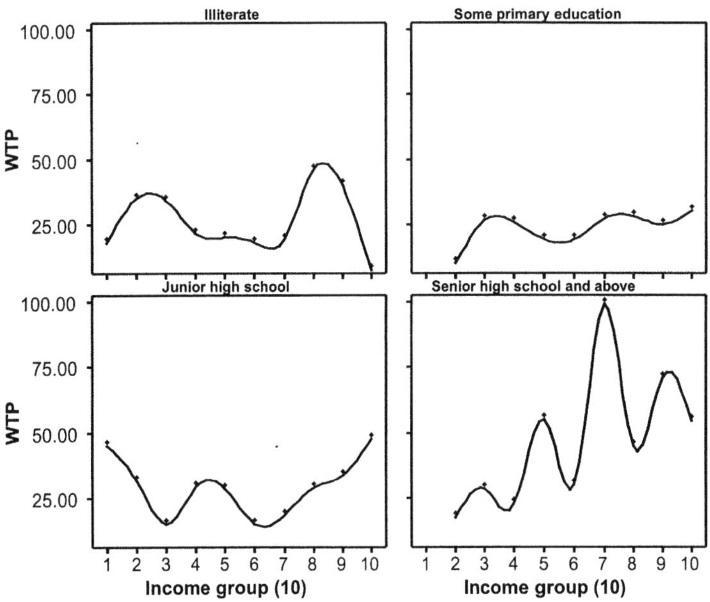

The variation of WTP with income when the effect of land area is controlled can be shown in this way. Likewise, households are divided into ten deciles in accordance with their income and the mean WTP of each decile is computed; then households are categorized into three groups in accordance with the scale of their land area: lowest one third, average and highest one third. Finally, the variation with income at different levels of land area is plotted in Figure 5-6. It is clearly shown that WTP spires with the increase in income by all the group of land area. Notice (1) that the smaller the land area is, the more the household is willing to pay; (2) that the poorest three deciles of households have little difference in WTP caused by different levels of land area. However, owning the lowest land area makes the average and richer households (the income deciles 4-10) willing to pay significantly more than owning the middle and highest land areas. One explanation is that though a household with high land area can produce more, the profit is not necessarily high; the large costs of agricultural production should subtract from total household income[21]. Another possibility is that psychologically the farmer really looks on the land as a type of security. After all, the tenure of land is tradable. Then, the substitutive effect between the land and CBHIS is present, especially when households are in the middle and upper groups in terms of total income. The similar tendency of variation of WTP for a complete coverage with *no* co-payment is observed when the effect of land area is controlled (Figure D-2 of Appendix D).

Figure 5-6 Variation of WTP with income at different level of land area

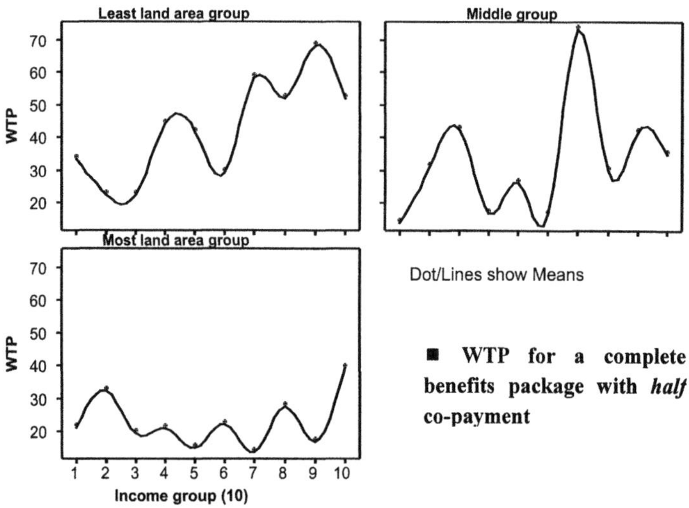

The variation of WTP for complete benefits package with half co-payment of two ethnic groups as income rises is plotted in panel (a) of Figure 5-7 in the similar way as stated before. An interesting phenomenon is found out that the relatively poor and the relatively rich deciles that belong to Yi nationality are willing to pay more than the corresponding deciles that are Han nationality. The

[21] Only 'total household income' is available in the dataset due to the complexity of calculating agricultural costs.

similar tendency of variation of WTP for a benefits package with no co-payment can be seen in panel (b) of Figure 5-7.

Figure 5-7 Variation of WTP with income at different ethnic group

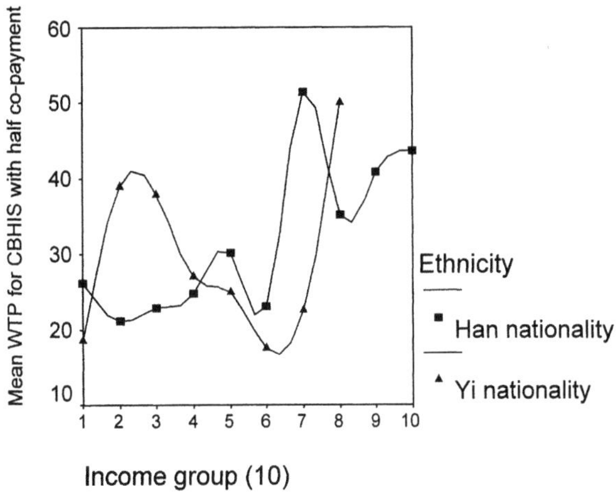

(a) WTP for a complete benefits package with half co-payment

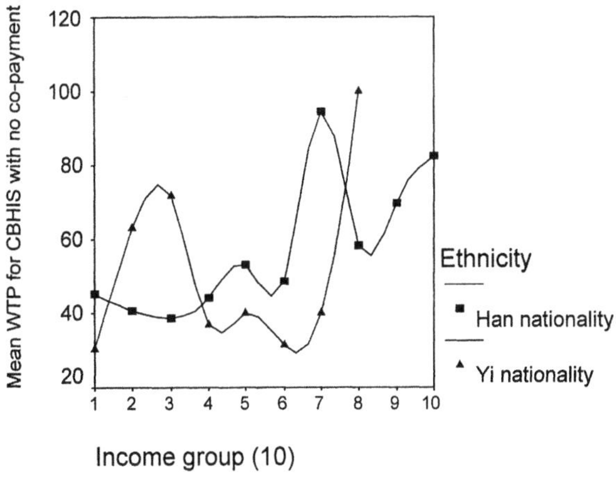

(b) WTP for a complete benefits package with no co-payment

Econometric analysis

Now that willingness to pay derived from the open-ended format can be treated as a continuous variable, a multiple regression model can be estimated through ordinary least square (OLS). The correlation matrix of all explanatory variables is employed to test their collinearity. One of any pair of variables that have perfect (R^2=1) or highly close ($R^2 \geq 0.7$) collinearity is excluded. The variables of WTP, income, and health care costs that are skewed are made a denary logarithmic transformation. The positive infinite value of the derived variables (e.g., ratio of dependants to independents) is replaced with the second maximal value. After the logarithmic transformation of dependent variable and a fraction of independent variables, the specification actually becomes a semi-logarithmic equation.

Figure 5-8 variation of WTP with income at different counties

- **WTP for a hypothetical complete benefits package with *half* co-payment**

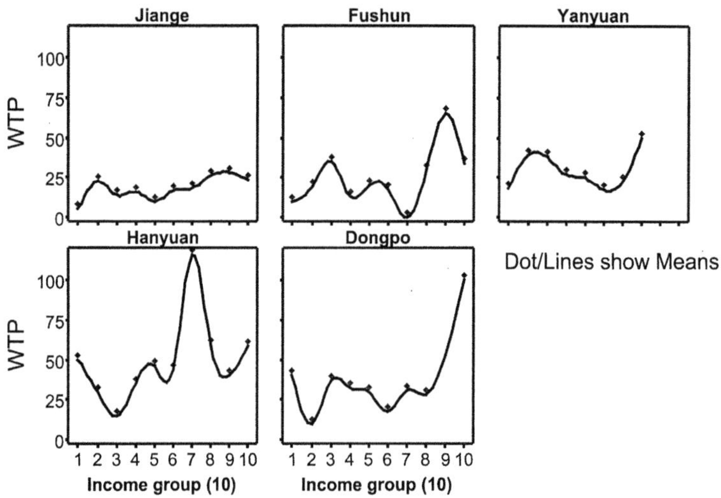

Estimation results of determinants of WTP for complete benefits package with no co-payment schemes and half co-payment are shown in Table 5-11 and Table 5-12, respectively. With respect to the former, two econometric models are specified. First, only household head attributes, household attributes, income, measurement of adverse selection and so on are allowed to enter the model, while the regional difference is ignored temporarily as shown in specification (1). Results indicate that seven variables are significantly related to WTP for non-co-payment scheme. Household cash income is positively related to WTP for the zero co-payment schemes. As pointed out in many researches using CV method to elicit the WTP, income is always a very important determinant. The coefficient of logarithm of cash income is 0.41, implying that WTP for the zero co-payment schemes increases by 0.41 percent as income rises by 1 percent, other conditions being constant. Mother's education is also a highly significant determinant. A one-year increase in mother's education brings about an increase in WTP of 6.1 percent, other conditions being held constant.

The governments of the developing countries are usually unable to provide social security for the rural population. Traditionally, it is viewed that the land has shielded the rural people against a fraction of risks. China is one of the advocators. Theoretically, the land serves as an asset, which can be turned into cash in time of need to deal with risk events. The success in liquidating land is to a large extent dependent on the developed land market. Nevertheless, the unfettered transaction of the agricultural land especially the cultivated land is legislatively prohibited in China. So the only role of agricultural land in risk management is its ability to generate income in the long run. The effect of the land on WTP is dependent on its capacity of generating profit. If the profit from agricultural production on the land is negative, then the amount of agricultural land has a negative impact on the WTP. The result is consistent with this theoretical prediction. An increase of one percent in land area reduces the WTP by 0.31 percent.

Ethnic group and whether a household has a migrant worker affect WTP significantly, i.e. two dummy variables. As pointed out previously, the variable of ethnic group is introduced into the model not only as the difference of sub-population but also as the difference of social capital in the community. Estimation result shows the ethnic group exercises a strong effect on the WTP. However, if the dependent variable is logarithmic, the relative change in mean *Y* for the dummy variable can only be obtained by taking the antilog (to base *ten*) of the coefficient of a dummy variable and subtract 1 from it (Gujarati 1995). Households in the Yi community are thus willing to pay 2.24 Yuan more than Han people, everything else held constant. Migrant workers are quite common for the rural households in Sichuan. The result shows a negative relationship between the WTP and the variable of whether a household has a migrant worker. Households having at least one migrant worker are willing to pay 0.48 Yuan less than those households not having migrant workers. The possible reason is the inconvenience of reimbursement and inability of RCMS to cover the relatively high costs of health care in case an illness occurs in costal cities; another reason is that the migrant workers in cities far from their hometowns have increased difficulty in monitoring the operation of RCMS, namely, they have more distrust in the scheme. Severe problems with the reconstruction of RCMS arise because of the high proportion of migrant workers in Sichuan Province. This may become the biggest obstruction to RCMS.

In addition, age of household head and distance of home from town center are both significant determinants. Age of household head is positively related to WTP. The coefficient indicates that a one-year-old increase causes WTP to increase by 1 percent.

There are still some insignificant variables worthwhile to discuss here. Adverse selection is a basic problem in insurance market due to the asymmetrical information between the insurer and the insured. It is believed that high-risk individuals value the insurance higher than low-risk ones, because they have a larger expected demand for health care. An attempt is also made to test the presence of adverse selection. Here last year's 'total health care cost' is used as a proxy. Estimation coefficients demonstrate that there seems to be slight adverse selection in terms of 'health care cost', but it is not statistically significant.

Table 5-11 Determinants of WTP for a RCMS with no co-payment

Dependent Variable: Log(WTP)	Coefficient	Standard Error	Coefficient	Standard error	Mean of X
Explanatory variables:	(1)		(2)		
Head attributes					
ETHNICT (1=minority)	0.526***	0.196			
HEADAGE	0.010*	0.005	0.010*	0.005	42.626
HEADEDU	0.007	0.017	0.010	0.016	5.504
SPOUSEDU	0.061***	0.017	0.047***	0.017	4.326
Household attributes					
HHSIZE	0.035	0.045	0.017	0.043	4.165
JOBBER (1=with migrant)	-0.169**	0.097	-0.114	0.095	0.452
DEP_IND	-0.042	0.084	-0.058	0.082	0.748
LG_DIST	0.220**	0.085	0.000	0.098	1.424
Income					
LG_LANDA	-0.309***	0.078	-0.072	0.099	1.825
LG_INCCS	0.407**	0.158	0.275*	0.159	3.756
Health risk magnitude					
LG_HLCST	-0.012	0.032	-0.016	0.031	1.770
CBHISEXP	-0.052	0.107			
Risk aversion					
CIGAR_FW	-0.180	0.123	-0.244**	0.119	0.613
CIGAR_MN	-0.065	0.154	-0.062	0.149	0.191
County					
CTY_JG			-0.291*	0.160	0.213
CTY_FS			-0.124	0.166	0.143
CTY_YY			0.313	0.220	0.226
CTY_HY			0.497***	0.172	0.209
Constant	1.811	0.676	2.311	0.638	
Number of observation	229		229		
R-squared	0.223		0.285		
Adjusted R-squared	0.172		0.231		
Log-Likelihood	-231.373		-222.339		
Restricted Log-Likelihood	-260.219		-260.880		
Model F test	4.38		5.30		
Durbin-Watson Statistic	1.981		2.066		

Source: computed from ZEF survey 2002

Since the decision to pay for a scheme needs to be made on a household basis, it is reasonable to test the impact of household's characteristics. The household size and the structure are thought of as a proper description besides the characteristics of household head and his or her spouse that have been addressed previously. Generally speaking, a big household has more income-generating resources and thereby a bigger risk pool in terms of risk management. For example, the agricultural land is evenly distributed to households on a capitation basis. Thus, a bigger household may be able to manage risk more readily than a smaller one. However, the household size does not fully convey sufficient information on capacity of income generation; another indicator, the capacity of income generation measured by ratio of independents to dependents, does more. The independent is counted on condition that one is at age from 18 to 65; the remaining is counted as dependents. The bigger the ratio is, the stronger and more stable they are capable of generating income, thus the better they cope with risks. We then hypothesized that a bigger household with a higher ratio of independents to dependents is willing to pay less (with a minus sign). Estimation results show that the regression coefficients bear positive signs, providing contradictory evidence to our hypothesis. The impact, however, is not only quite slight but also has no statistical significance.

Second, the county difference is taken into account to see if the large variation in the above variables comes from different counties and how large the marginal effect of willingness to pay is. We have 4 dummies of counties but due to the problem of high collinearity between the ethnicity and RCMS experience versus the county dummies, the dummy variables of ethnic group and RCMS experience are kicked out in the OLS estimation. Estimate results demonstrate that the WTP in county Jiange is significantly lower than the reference county, Dongpo. The willingness to pay of Hanyuan households is significantly higher than the reference county. The other two variables of county are not statistically significant.

After county dummies enter the model, all the independent variables bear the same signs as in estimation (1) and only the magnitude and significance lower a bit. Three of the significant variables become insignificant, which are the dummy of migrant worker (JOBBER), distance (LG_DIST) and land area (LG_LANDA). This suggests that variation of JOBBER, DISTANCE and LANDAREA are largely accounted for by regional difference. Also, one insignificant variable, cigarette smoking (CIGAR_FW), turns significant.

F tests verify that each of the two models is valid in general and these explanatory variables do help explain the variation of WTP about its mean. Goodness-of-fit can be shown by R square. The R-squared values are 0.22 in estimation (1) and 0.29 in estimation (2), indicating that 22% and 29% of variation of WTP is explained by the variables included in the regressions.

The determinants of WTP for the hypothetical RCMS with a co-insurance rate of 50% are almost the same as those for RCMS with no co-payment, whether the significance or the sign (see Table 5-12). Regarding the estimation without county variables, significant variables are one fewer. Whether a household has a migrant worker is not a significant determinant any more. This suggests that households with migrant workers may prefer a less generous coverage to a generous one (coinsurance rate=0). A useful policy implication is that for a community that has a large proportion of households with migrant workers, RCMS with half coverage is preferable to that of full coverage. When county dummies are introduced, distance and land area turn insignificant, while some of the county dummies turn significant.

This also suggests that variation of JOBBER, DISTANCE and LANDAREA are largely accounted for by regional difference. In addition, in comparison with the reference county of Dongpo, households in Hanyuan and Yanyuan are significantly willing to pay more, while households in Jiange are significantly willing to pay less for the hypothetical RCMS with half co-payment.

Table 5-12 Determinants of WTP for a RCMS with 50% co-payment

Dependent variable: Log(WTP)	Coefficient	St. Error	Mean of X	Coefficient	St. Error	Mean of X
Explanatory variables	(3)			(4)		
Head attributes						
ETHNICT (1=minority)	0.728***	0.208	0.227			
HEADAGE	0.012**	0.006	42.681	0.011**	0.005	42.626
HEADEDU	-0.001	0.018	5.493	0.002	0.018	5.504
SPOUSEDU	0.080***	0.018	4.306	0.068***	0.018	4.326
Household attributes						
HHSIZE	0.032	0.048	4.170	0.019	0.047	4.165
JOBBER (1=with migrant)	-0.137	0.103	0.450	-0.091	0.103	0.452
DEP_IND	-0.067	0.090	0.752	-0.088	0.089	0.748
LG_DIST	0.245***	0.090	1.424	0.064	0.106	1.424
Income						
LG_LANDA	-0.310***	0.083	1.822	-0.089	0.107	1.825
LG_INCCS	0.381**	0.168	3.754	0.255*	0.172	3.756
Health risk magnitude						
LG_HLCST	-0.003	0.035	1.778	-0.008	0.034	1.770
CBHISEXP	-0.061	0.114	1.563			
Risk attitude						
CIGAR_FW	-0.090	0.131	0.616	-0.150	0.129	0.613
CIGAR_MN	-0.015	0.164	0.192	0.001	0.161	0.191
Counties						
CTY_JG				-0.318*	0.173	0.213
CTY_FS				-0.060	0.179	0.143
CTY_YY				0.496**	0.238	0.226
CTY_HY				0.406**	0.186	0.209
Constant	1.084	0.719		1.507	0.689	
Number of observation	229			230		
R-squared	0.210			0.251		
Adjusted R-squared	0.158			0.195		
Log-Likelihood	-245.75			-240.09		
Restricted Log-Likelihood	-272.67			-273.37		
Model F test	4.05			4.47		
Durbin-Watson Statistic	1.888			1.929		

Source: computed from ZEF survey 2002

5.2.4 Validity test of CV results

Most studies construct an equation that predicts WTP for a good as a function of several other variables in surveys, such as income, past use, and various attitude and knowledge concerning the good in question. An equation having reasonable explanatory power and the expected signs provides evidence in support of the proposition that the survey has measured the intended construct (Carson 2000). In accordance with this criterion, the results of CV survey in this study is quite reliable because the overwhelming majority of explanatory variables bear the expected signs, especially the key variables such as income, education, migrant worker, whether in the estimations of willingness to join or willingness to pay.

There are two simple methods to test the validity of CV results: price and scope tests (Carson 2000; Carson *et al* 2000). The first one refers to a well-known economic maxim: the higher the cost, the lower the demand. In the binary discrete choice format, this can be easily tested. In the open-ended format under consideration, unfortunately, this law cannot be tested. The scope test refers to testing the law that respondents are willing to pay more for a larger amount of desired good. Our format design can test this law readily. As shown in Figure 5-3, for a complete coverage (out- and inpatient) with half co-payment the respondents are willing to pay more than for inpatient services coverage while willing to pay less for a complete coverage with no co-payment. The success in the scope test demonstrates the validity and reliability of CV results based on an open-ended question format. Despite the economic consistency and validity of the CV results, it is reported that the estimated consumer surplus per permit in real market was lower than the willingness to pay in the hypothetical market after comparing actions in hypothetical hunting-permit market with a real market (Bishop and Heberlein 1979). We should be cautious about using the mean of WTP or median WTP as a criterion of planned premium level. A premium that is a bit lower than the stated mean of WTP is feasible and reliable.

5.3 Summary and recommendation

Literatures reviews and our case study indicate that rural cooperative medical scheme in China has different capacity of resources mobilization and risk protection in different regions. The Eastern developed region not only has a higher enrolment rate but also mobilize more resources from governments and local communities and hence protect their members better against costs of illness. By contrast, middle and western regions have low enrolment rate, weak capacity of resource mobilization and risk protection. Attempt was also made to link performance of RCMS to health improvement. The comparative analyses show that RCMS as a whole played a big role before the economic reforms in mobilizing resources and hence enhancing the national health status. However, after the economic reforms, RCMS appears to have no significant role in enhancing the health status at the county level due to the low coverage. This preliminary conclusion needs further examination.

To make RCMS more desirable, an improved RCMS was hypothesized and willingness to join (WTJ) and to pay (WTP) was investigated among 300 sampled rural households in 10 villages in Sichuan province at the beginning of

2002. Findings show that willingness to join (69%) hypothesized RCMS is not high in consideration of the high enrolment rate (over 90%) of RCMS in the 1960s and 1970s. But the percentage of WTJ is high sufficiently to sustain the operation of RCMS in a town. Among three hypothetical benefits packages of RCMS are inpatient services coverage with no co-payment, complete coverage with half co-payment and complete coverage with no co-payments. The second benefits package is most desirable because WTP for it exceeds half the WTP for complete benefits package with no co-payment. Although WTP for hypothetical RCMS with half co-payment has a reimbursement ratio similar to the present RCMS, the mean of WTP (32.50 Yuan) is much higher than the premium of the present RCMS (6 Yuan). The hypothetical RCMS with half co-payment distinguishes itself from the existing RCMS by its strict insurance fund management and enlarged benefits (outpatient services also covered). Thus, the capacity of risk protection and strict fund management is crucial to enhancing WTP. Moreover, the regression analyses show that the attributes of household head (age, mother's education and ethnic group), household attributes (distance from center, whether with migrant worker) as well as income are closely related to WTJ and WTP, of which being minor ethnic group and having higher income are two strong boosting factors whilst having a migrant worker is a significantly negative determinant.

The survey results and econometric analysis suggest that governments and communities need to enlarge the capacity of RCMS to mobilize resources and to protect against costs of illness. To fulfill this task, RCMS should be designed to be a complete benefits package with a coinsurance rate between zero and 50%, private contribution of about 30 Yuan per capita per year, strict management of insurance funds and enhanced the level of subsidies from local and central governments, especially for rural people in the underdeveloped provinces in Middle and Western regions. In the process of reviving RCMS, the central, provincial and county governments should play an active role in subsidizing, guiding and regulating; the town government and village should be major executive agencies and administrators. Furthermore, the minority communities with middle income can be prioritized if they have possessed the basic capacity of implementing and managing RCMS. However, the high proportion of migrant workers is a great barrier to the establishment of RCMS, it could be weakened or removed by adopting a half co-payment scheme.

6. Strategies for Coping with Costs of Illness and the Role of Extended Family in Risk Coping

The previous two chapters addressed demand for health insurance (private health insurance and rural cooperative medical schemes). In the process of analyzing demand for health insurance, it is implicitly assumed that the insurance contract is the only available asset for hedging risk. In fact, the insurance contract is not the only strategy of risk management; it is only a subset of assets of an individual's portfolio. Demand for health insurance and demand for other assets in the portfolio are determined simultaneously (Mayers *et al* 1983). In a perfect capital market framework there is no demand for a specific insurance contract even though risk aversion and uncertain future consumption are explicitly assumed. Individuals can sell shares in any claim (including human capital claims) and effectively eliminate insurable risks through diversification. Thus, a necessary condition for a specific demand for insurance is that the cost of eliminating risks through diversification exceeds the cost of hedging them with insurance (Mayers *et al*1983).

In rural areas of developing countries, on the one hand, both the capital and insurance markets are underdeveloped or missing. On the other hand, the magnitude and the time when illness costs occur are uncertain, exposing rural people to the risk of consumption. In addition to risk of consumption, the rural poor are less well insured in income (Jalan *et al* 1999), which in turn makes consumption more uninsured. Without formal insurance arrangements in most rural areas, there must be some informal risk sharing arrangements that can be utilized to cope with a wide variety of risks. A large number of strategies are observed in rural areas of low-income and developing countries. For instance, the rural households not only try to diversify their income, but also adopt strategies such as paying cash, drawing on savings, selling assets, borrowing money, exchanging gifts or money, or even letting the school kids drop out of school. With these strategies, various risks can be coped with at the levels of the household, extended family or community (e.g., Sauerborn *et al* 1996a; von Braun *et al* 1999; Morduch 1999; Zeller 1999; Weinberger *et al* 2000; Gertler and Gruber 2000). These strategies have been effective in many parts of the world and many times of history. The existing strategies must have some rationality. However, these coping strategies, particularly the risk sharing arrangement within the extended family, have not been paid due attention. This chapter intends to analyze health risk coping strategies in general, and the risk sharing arrangement within the extended family in particular. Furthermore, the informal insurance arrangement (the extended family) is carefully compared with formal insurance arrangements (private health insurance and RCMS) in terms of institution, organization and management.

6.1 Costs of illness and household coping strategies in rural China

6.1.1 Conceptual framework and theory

The costs of illness could be substantial if both the economic and non-economic costs are taken into account. Non-economic costs such as the decrease in utility from other consumption, pain, and even the worry about premature death are immeasurable. Even in economic terms, the indirect costs of illness are difficult to measure. For instance, the illness may reduce productive capacity, or persons infected by some communicable diseases (or being non-symptom carriers) may be excluded from labor markets. The exclusion from the labor market would undermine the capacity of income generation of the ill permanently. However, in this chapter we are mainly concerned about the direct costs of illness, with which household have to cope. The most proximate direct costs of illness for individuals and their households include the costs of treatment and lost work time (Hamoudi and Sachs 1999). Mead (1992) estimated the average total costs of treatment and foregone productivity in Tanzania resulting from a single HIV infection to be about $2462-5316 in 1985 dollars, or about 8.5-18.3% of per capita income. Mild costs of illness can be financed with cash holding or deposits. However, when the costs of illness are large relative to income or savings, individuals or their households have to adopt other strategies. Whether or not coping is needed depends on the ratio of costs of illness to total household income. Observation indicates that people do not passively cope with costs of illness. Rather, they actively adopt measures to reduce risk and the loss of risk, that is, they do manage the risk. Even if the household's annual income is sufficiently high to cover the costs of illness, the seasonal fluctuation of the costs of illness is not necessarily synchronous with the fluctuation of earning money (Sauerborn *et al* 1996b). Thus, risk coping remains needed for these strata of households.

The theory of risk management on a corporation basis explicates that the value of corporation is determined by the magnitude and time of future net cash flow and its risk. The occurrence of a variety of risk events will cause the unexpected fluctuation of cash flow that then accordingly leads to the fluctuation of share price. The existence of risk basically depreciates a corporation and enhances the expected return for investment of shareholders. Therefore, the objective of risk management is to maximize the value of corporation by minimizing risk cost. Risk cost can be defined as the decreased value of corporation due to the existence of risk (only pure risk is considered for convenience). Risk cost generally is composed of five integral components: expected cost of loss, cost of loss control, cost of loss financing, cost of internal risk reduction and cost of residual uncertainty.

Accordingly, for an individual or a household to manage health risk in the context of rural areas in low-income countries, this theory remains an important reference. An individual manages her health risk in order to maximize her wealth (welfare) by minimizing health risk. Theoretically, the cost of health risk can also be classified into five types, like the cost of corporate risk. Empirically, the division of health risk cost is neither economical nor realistic. However, the classification is helpful for understanding costs of health risk and then minimizing them.

Risk management is defined as a systematic process for the identification and evaluation of pure loss exposures faced by an organization or individual, and for selection and implementation of the most appropriate techniques for treating such exposures (Rejda 1997). To deal with issues of risk management in modern corporation, Rejda proposes five techniques for treating loss exposure: avoidance, loss control, retention, non-insurance transfers and insurance. This definition of risk management is applicable to the risk management of rural households or individuals, and the corresponding techniques are useful for guiding risk management of rural households. However, these techniques are not quite practical for the rural households owing to their different institutions, organizations, the environment they reside in and the pattern of risk exposure from the modern corporations. Some economists summarized the risk management techniques specific for the rural households (e.g., Zeller 1999; Holzmann *et al* 2000). Based on the compilation of Weinberger *et al* (2000), the risk management of rural households can be divided into three categories: risk reduction, risk mitigation, and risk coping. This thesis follows the latter typology because the study was conducted in rural areas. Though some attention is paid to risk reduction and mitigation, the focus of this section is put on the risk coping strategies.

Risk reduction

A common way to deal with risk is simply to reduce risk itself. Risk reduction refers to techniques or strategies that lower *ex ante* probability of a risk event. Reducing risk of health can be achieved in a wide variety of ways by multiple levels of organizations, all of which imply altering health production choices. First of all, people reduce health risk by selecting and modifying the environment. The type of settlement in some parts of Sub-Saharan African is a case in point; people tend to locate themselves in the Eastern African highlands because malaria risk is lower at high altitudes. The ancient society of China provides another example. When the contagious diseases were not controlled, the rural houses were built very dispersedly in the hilly or mountainous areas to avoid cross infection. Even now when most of severe contagious diseases of human beings have been controlled throughout the country, this convention is still maintained since a lot of poultry and husbandry animals are prone to such contagious diseases as avian influenza, or foot and mouth disease.

Health risk, in the final analysis, is chance of financial loss, thus any form of techniques and strategies to stabilize income streams somewhat belong to the category of risk reduction. Rural inhabitants often seek to minimize their exposure to risk by diversifying their portfolio of income generating activities, by self-sufficiency and by flexibility and so on (Fafchamps 1999). In areas with less extreme climatic conditions, for instance, farmers often plant different crops, or several varieties of the same crop to obtain a more stable yield. Intercropping is often partially justified by the consideration of risk management.

Not only rural households attempt to reduce health risk, but also local communities and governments of developing countries are deeply involved in risk reduction by improving the health infrastructure and providing health education. For instance, programs aimed at securing drinking water and building sanitary toilets have been successfully undertaken in rural China. An increasing number of rural people obtain access to clean and safe drinking water and sanitary toilets. Another strategy that the government has taken is to provide preventive health care through its public health system, which may decrease both cost of loss control of individual household and the social cost of risk. Health education is an effective way of spreading the knowledge of health care. With the sufficient knowledge on health maintenance, people can reduce health risk through a healthy way of life, scientific diets and so on. However, the success in effectively reducing risk needs the common efforts of households, communities as well as the governments.

Risk mitigation

Risk mitigation represents the techniques and strategies that mitigate the loss when risk events take place (Rejda 1997). A fire extinguisher is a case in point. Extinguishers do not reduce the incidence of fire, but they do mitigate the severity of fire. The major loss of health risk is financial loss and consumption fluctuation caused by financial loss. For health risk the 'extinguisher' can be timely and proper health intervention, which greatly mitigate the consequence of illness. The 'extinguisher' can also be investments and accumulation of assets as buffer stock to smooth consumption. More specifically, the severity of loss caused by illness can be mitigated via delivery of social and financial services, e.g., credits, savings, insurances, educations, nutrition and health.

Risk coping

Risk coping is defined as a short-term strategy adopted within the prevailing value system to avert a negative effect on the actor (Davis 1993). However, the boundary between risk mitigation and risk coping is relatively ambiguous. It can be illustrated by the example of the fire extinguisher as well. The purchase, installment and maintenance of extinguisher belong to risk-mitigating activities. Once people open the extinguisher to put out the fire, this behavior is falling into the category of risk coping. By the same token, the process of saving in the form of asset, for example, is to mitigate the consequence of health risk for rural people who have little access to financial services. But when one liquidates assets in asset market in order to pay the bill of health care in case sickness occurs, this is risk coping. The distinction between the two is that the risk mitigation is ex-ante behavior while risk coping refers to ex-post activities.

Asset liquidation is likely to have a negative impact on future earnings. Distress sales of productive assets such as farm livestock, oxen, farm tools, vehicles and farm building, for example, may cause the household to undergo a long-term fall in production. Precautionary savings in the form of money is a way in which households can protect themselves against the damaging consequence of distress sales of productive assets: in anticipation of future shocks, households may build up liquid reserves that can be sold or consumed in times of need. The liquid assets that are accumulated in this manner by poor rural dwellers include food stocks, gold and jewelry, cash, deposits and checking accounts (Lim *et al* 1994; Behrman *et al* 1997). Faced with a sufficiently long series of bad income and utility shocks, households will nevertheless run out of assets and will no longer be able to absorb shocks by liquidating assets. One conceivable way out of this dilemma is to let households' assets position become negative, that is, to let households borrow. As Carroll (1992) has shown, however, credit constraints are unavoidable if credit contracts are strict, that is, if the creditors insist on repayments under any circumstance.

The emphasis in this section is placed on *ex-post* risk coping. The actor of risk coping is the household defined as a group of individuals—most commonly but not necessarily linked by kinship ties—who reside together and share functions of production and consumption as well as of reproduction. The negative effect that coping strategies seek to avert is the breakdown of the household as an economic and social entity. In the case of a famine research, the criteria for calamity are household destitution and disintegration (Sauerborn *et al* 1996a). Household responses involve trade-offs between and within various coping options. In other words, different households within the same community stand at different points along the continuum, and their responses to a given threat will vary according to their endowment base, access to community support, and access to public intervention (von Braun *et al* 1999).

Table 6-1 Possible *Ex post* Methods of coping with illness-risk

Level	Household	Extended family	Above extended family	Alternative
Financial costs	Cash, withdrawing savings, sale of assets, wage-labor, sale of consumer durables	Gift-transfer, cash-transfer, free interest borrowing, lower-cost service searching (using social capital)	On trust by private doctor, free care, gift-transfer within local community, government relief, loans from financial institutions or money-lenders, and drop-out	Most can be replaced by PHI policies, RCMS
Time costs	Intra-household labor substitution, changing the capital-labor mix of production	Labor transfers for nursing or farm work	Hiring labors, free community labors	Alternative unavailable

Source: own compilation

Besides the category of risk management strategies, the level at which the response takes place is important. Although there are many levels of organization that may be involved in health risk coping, this section mainly undertakes analyses at the levels of household, extended family and above extended family. Most strategies are aimed at the financial costs of illness, while the strategies aiming at time cost are somewhat neglected. Nevertheless, for rural households who make a living mainly by farm labor or wage labor, time costs are not negligible. As shown in Table 6-1, four-level strategies for coping with financial costs and time costs of illness are summarized, including formal health insurances as a reference.

6.1.2 Methodology and data

Many studies on how the poor cope with various risks and hazards have summarized a series of coping strategies that the poor households adopted. Chen's (1991) narrative of a drought-affected village in India contains a good summary of the coping strategies of the poor: "growing a mix of crops and/or rearing a variety of livestock, entering the labor and tenancy (sharecropping) markets as needed, drawing down stored goods or fixed assets, adjusting consumption, borrowing, using common property resources, migrating (seasonal), and drawing upon traditional social security arrangements." Sauerborn *et al* (1996a) has examines the strategies of rural households in Burkina Faso to cope with costs of illness. A large inventory of coping strategies has also been offered in detail therein.

Despite the large inventory of coping strategies, some new strategies for coping with the economic cost of illness (especially with time costs of illness) in the context of rural China are added. Applying an approach modified from Chen (1991) and Sauerborn (1996a), the researcher analyzes strategies for coping with economic costs of illness in rural Sichuan along four dimensions as follows.

- The coping behaviors are described. Based on the classification of coping strategies in Table 6-1, we designed some indicators in the questionnaire to describe rural households' risk coping behaviors.
- The levels at which health risk is coped with are identified. Coping may be confined to the household in which illness occurs, or it may involve the relatives or non-kin friends within the extended family. All the coping strategies beyond the extended family are considered to be at the third level.
- The sequence of strategies adoption by households is analyzed.
- A judgment of success in coping is made. Three economic criteria are used to judge whether household risk coping is successful: First, did coping avert calamity—i.e. the social and economic breakdown of the household? Second, did it reduce or avert any negative effect of illness on household production? Third, did coping strategies preserve household assets?

This particular analysis is again based on the Health Insurance Survey in Rural Sichuan (HISRS). For the analysis, quantitative indicators were designed and surveyed, which include the following.

- Household size and ratio of dependants are utilized to test a hypothesis that strategies in large households and in households with a low dependency ratio are more effective in coping with health risk.
- Household wealth is utilized to test a hypothesis that, though the wealthy households have a greater arsenal of coping strategies than the poorer households, they succeed in coping health risk by using fewer strategies.
- Age, gender and relation to household head of sick individual are utilized to test the hypothesis that the allocation of household health resources is biased towards the productive, male members and kids of the household heads.

Additionally, the direct costs of illness[22], has two integral components.

- Financial costs of illness[23] include user fees and other expenses. User fees refer to the major part of financial costs charged by providers for the treatment or services concerned. This expenditure, however, cannot represent all the financial costs of sickness. Expenses of round traveling, board and lodging if the provider is far or the ill has to be hospitalized ought to be taken into account too. 'Other expenses' refer to the expenses occurred for traveling and foods of the patients, and the lodging of companion[24]. We hypothesize that the larger the aggregate financial costs of sickness, the more strategies the household adopts.
- Time costs of illness deserve examination apart from financial costs, especially for the productive members in agricultural productive season. Days lost to illness both by the sick and by the healthy members who spend time tending the sick and/or accompanying them to a distant treatment site are two indicators. Time costs were computed by converting days off work due to illness into monetary terms at the temporal daily wage rate for rural laborers, which provides a proxy for the opportunity cost of time.

The enumerators interviewed household heads or household wives if household heads were unavailable (e.g., they migrated to other places as casual workers) based on a questionnaire. The data on individuals reporting sickness in the past year were collected. At first thought, one year may be a bit long for respondents to recall the expenditures exactly. However, except normal productive and living expenditures, rural households have few items of expenditure that are unexpected, especially among the poor rural. Thus, the data are assumed to be reliable and valid.

6.1.3 Empirical Results

6.1.3.1 Direct costs of illness

Table 6-2 presents the main components of a household's direct costs of illness:

---

22 Indirect costs of illness are excluded from the quantitative analysis for its immeasurability.
23 Financial costs of illness and medical costs are used in the chapter interchangeably.
24 The lodging fees of the patients are included in the user fees.

financial and time costs. The county of Jiange has on average the highest user fees while the Dongpo district has the lowest. It can also be seen that the medical cost per household ranks in the same sequence as user fees. It is striking to note that the direct costs of illness varies so considerably that the highest is almost 5 times as large as the lowest, either in terms of financial costs or in terms of time costs. Even though the large variation can be explained by many social and economic factors, the difference in propensity to report illness may have a major effect. For instance, people in Jiange seem to report severe illness only and ignore a minor ailment and hence the average direct cost of illness is larger than that of others.

Table 6-2 Components of direct costs of illness per household per year

	Jiange	Fushun	Yanyuan	Hanyuan	Dongpo	Total
Financial costs of illness (Yuan/household)	4884.09	1057.42	1185.14	3346.25	914.2	2020.20
User fees	4227.27	912.57	1023.17	3076.83	815.72	1838.30
Other expenses	656.82	144.85	161.97	269.42	98.48	181.90
Time costs of illness (Day/household)	331.99	24.76	180.71	87.13	25.93	108.90
Time unable to work	218.24	15.83	74.14	33.75	4.65	53.60
Accompanying time	64.95	5.14	96.28	17.68	19.35	37.60
Treatment time	48.8	3.79	10.29	35.7	1.93	17.70

Source: ZEF Survey 2002

Table 6-3 shows the percentage of the components of the direct costs of illness. The share of time costs in the direct costs of illness varies largely across counties, from 28.1% to 69.6%. The large variation in time costs may be accounted for by the uniform wage rate. Due to the lack of local average wage rate, a general average wage rate of rural labors at the province level is utilized for conversion, which, however, may over- or underestimate time costs. For instance, the wage rate in Yanyuan that is remote and exceedingly underdeveloped may be overestimated and hence time costs are artificially elevated; on the other hand, the *de facto* low wage rate gives people less incentives to treat disease out of the consideration for labor loss, compared with those in other sampled counties. Hence medical costs are lowered. Alternatively, the large variation in share of time costs in the direct costs of illness can be attributed to different medical care price indices at different levels of health facilities. As shown in chapter 4, the majority of people in Yanyuan seek care in village health stations where health care is much cheaper. Meanwhile, the majority of Jiange people seek care in town health centers that provide health care of high quality and price, relative to that of village health stations.

The share of direct costs of illness in income per household is an indicator reflecting the affordability for direct costs of illness. In the last row of Table 6-3, average income per household in different counties and the share of direct medical costs in average income are calculated. As a result, the proportion of illness burden to income varies largely across counties, from 5.4% to 47.3%.

Direct costs of illness take up 29.0% of average income per household when taking the sampled households as a whole.

Table 6-3 Percentage of the total costs of illness Unit: Yuan per household

Type of illness costs	Jiange		Fushun		Yanyuan		Hanyuan		Dongpo		Total	
	Amount	%	Amount	%	Amount	%	Amount	%	Amount	%	Amount	%
Time costs	4979.9	50.5	371.4	26	2710.7	69.6	1307	28.1	389	30	1633	44.7
Financial costs	4884.1	49.5	1057.4	74	1185.1	30.4	3346.3	71.9	914.2	70	2020	55.3
Direct costs of illness	9863.9	100	1428.8	100	3895.8	100	4653.2	100	1303	100	3653.2	100
Income	10318	47.3	14875	7.1	7186	16.5	13466.7	24.8	17055	5.4	12580	29.0

Source: computed from ZEF survey 2002

Note: costs of time are computed by multiplying days lost to illness by average daily wage rate in Sichuan Province, 15 Yuan per day.

For the purpose of an international comparison as made in the section of discussion, the annual direct costs of illness per household is converted into monthly direct costs of illness per sick person by dividing it by 12 months and by average household size. The unit of Chinese Yuan is exchanged for US Dollar according to temporal exchange rate (Table 6-4).

Table 6-4 Monthly direct costs of illness per sick person in US dollar

	Jiange	Fushun	Yanyuan	Hanyuan	Dongpo	Total
Time costs (US$)	10.03	0.84	5.82	3.28	0.98	4.19
Financial costs (US$)	9.84	2.40	2.54	8.40	2.31	5.10
Direct costs of illness (US$)	19.86	3.24	8.36	11.68	3.29	9.29

Source: computed from ZEF survey 2002

Note: Exchange rate of US dollar for Chinese Yuan is 1 US$=8.3 Yuan at the end of 2001.

6.1.3.2 Types of coping strategies

Our taxonomy of coping behavior comprises 10 strategies. All of them cope with financial costs, some of them can also cope with time costs directly. As shown in Table 6-5, of 252 illness episodes, 101 (or 40.1%) are successfully financed through only one round of endeavor, 73 (or 29.0%) through 2 rounds of efforts and the remaining 15.5% through 3 rounds. However, we cannot take it for granted that those who made only one or two rounds of endeavor had raised sufficient resources for health care, they may be under-treated due to the insufficient solvency. The coping strategies are analyzed as follows.

Using cash and drawing on savings is a fundamental coping strategy. The result illustrates that some 86 percent of illness episodes, in the first place, are financed by using cash on hand and by drawing money from a bank account. It also indicates that the overwhelming majority of households have cash holdings or a bank account to which they can resort. However, there are still 35 episodes of illness (or 14%) that are financed direct through other channels. Put it differently, these households have no cash on hand or no deposit balance on the bank account (or even no bank account at all) to settle the medical bill as household members fall sick. These households are compelled to adopt other strategies to cope with the costs of illness.

Sale of poultry and livestock is the second most important in the first round of health care financing. Some 4 percent of all cases report the immediate use of this method due to the lack of cash or deposits. Sale of poultry and livestock plays a more marked role in the second round of financing. As shown in Table 6-5, of 112 episodes of illness that need the second financing, 46 (or 41%) have to rely on this method after households failed to pay off the medical bills in the first round. As a matter of fact, chicken and swine are reared most commonly in rural Sichuan, 77% and 70% of sample households reported rearing at least one swine and one chicken, respectively. The rationality of these farm activities is that, besides the two animals' function as a source of income, chicken, a small domestic animal, can serve as a current deposit that the farmers may draw quite handily and flexibly and help deal with a small amount of expenses. The swine serves as a fixed deposit to cope with costs of severe sickness. Similar reports in African countries described the two husbandry animals as "ambulatory savings-bank" (e.g., Sauerborn *et al* 1996a).

Sale of other assets is a method similar to the previous one. Three out of 252 episodes reported this method as first step, six out of 112 reported this as the second and three out of 39 have reported it as the last resort. Although this is still an effective strategy for coping with the costs of illness, it should be looked on as a failure of risk coping because the sales of assets may lower the productive forces (in case of the sale of productive assets) and the level of consumption (if consumptive assets are sold).

Loans from relatives play an important role similar to sale of poultry and livestock in the first round. Some 4 percent of all episodes are immediately paid for through the money borrowed from relatives. Around 23 percent of episodes that need the second round of financing settled the medical bill with the aid of loans from relatives. Generally speaking, loans from relatives are interest free. In contrast, the moneylenders or financial institutions must charge interests for the loan. Particularly, the moneylenders charge an interest rate that is a bit higher than the official interest rate (however, it is not as high as that of the usury, which is illegal in China). The loans from moneylenders are quite undesirable for two reasons. First, the borrowers are imposed an extra economic burden because not only is the interest rate high, but it is also computed by means of compound rate. Second, loans from moneylenders have to be paid back under any circumstances. In case one borrower fails to return the money as scheduled, the lender might compel the borrower to return the loans via threat of using illegal forces[25].

---

[25] Sources from the rural financial intermediates (e.g., Rural Credit Cooperative) indicate that pure consumptive loans have already been available since the end of 2001, but few applied for loans because of the collateral requirements. In 2002, the rural credit cooperatives were

Table 6-5 Sequential coping strategies of financial costs of illness

	1st financing		2nd financing		3rd financing	
	Frequency	%	Frequency	%	Frequency	%
Own cash or savings	217	86.1	0	0	0	0
Sale of livestock	11	4.4	46	41.1	3	7.7
Sale of other assets	3	1.2	6	5.4	3	7.7
Loan from relatives	11	4.4	26	23.2	3	7.7
Transfer from relatives	2	0.8	16	14.3	19	48.7
Loan from money lender	0	0	1	0.9	2	5.1
Allowing for payment delay	2	0.8	1	0.9	1	2.6
Free treatment	2	0.8	1	0.9	0	0
Others 1 (transfer from offspring)	4	1.6	5	4.5	1	2.6
Others 2 (reimbursement by insurers)	0	0	10	8.9	7	17.9
Total	252	100.0	112	100.0	39	100

Source: ZEF survey 2002

Transfer from relatives is rare in the first round of risk coping. However, it is important in second round of coping. Some 14 percent of cases that had to be covered by adopting the second method, received private transfer in the forms of money and gift from relatives. The transfers from relatives play the most leading part in risk coping as the last resort. Some 48 percent of those who used a third method acquired the financial aid from their relatives, whether in the forms of gift and money, or in the form of labor. The transfers from relatives can also deal with the time costs by transferring labor to the household with a sick member. However, the sick member or another household member would render back a corresponding amount of labors after the recovery on the principle of reciprocity.

Allowing payment delay for health care is only observed with private health care providers. This does not mean that private providers are more generous or charitable than public ones. It is the fierce competition on the market for health care that compels the disadvantaged small-sized private clinics to adopt this approach. However, it is not common. Altogether, there are only 4 illness episodes being allowed a grace time for due payment, since these poor households were in financial trouble at the time of illness.

Free treatment is another way of coping with health care expenses for the poorest of the poor. Of all the cases, three received offers of free treatment. Free treatment becomes possible when three conditions are met. One is that the user fee is low; another is that the insolvency of the sick is self-evident (e.g., the childless aged); finally the sympathizer is a private or a witch doctor.

---

moving forward a big step. They launched a great project of credit examination of a portion of rural households. Loans without any collateral become possible now for those who have already passed the credit examination.

The last coping strategy "*others*" includes two methods that had not occurred to the researcher before the formal survey: one is that the sick parents received obligatory transfers from their offspring; the other is that the insured got the reimbursement from an insurer or administrator of a rural cooperative medical scheme. Altogether there are 9 sick persons who received obligatory transfers from their sons or daughters. For the latter, in total 17 persons had part of their health care expenditure reimbursed. According to the procedures of insurance companies or RCMS, the patients need to settle the medical bill out of pocket first and then have it repaid. Thus, the reimbursement of the insurance company or RCMS generally acts as the second or third round of coping strategy.

6.1.3.3 Sequence of strategies adoption

To explore the sequence of adopting health risk coping strategies, the respondents were requested to orderly come out with the coping strategies they used and the interviewer wrote down three strategies. Statistics reveal that 95% of illness events were coped with within three rounds. As the first step, households utilized 7 out of 8 strategies (except loan from money lenders) that had been listed on the menu. Equally 7 strategies were used as the second financing, excluding "own cash or savings"; the third strategies concentrated solely on 6 means. Two of all eight strategies, "own cash or savings" and "transfer from relatives", were the most remarkable. Cash or current deposit was certainly the first used strategy for coping with financial costs of illness and no one used it as second or third strategy. Transfers from the relatives could be used in any round, but most frequently it was used as the last resort (Table 6-5). The sale of non-productive livestock such as matured poultry and swine is the second important option in the first round, and sometimes the rural households also sold some excessive farm animals or the calfhood animals. To sum up, households adopted coping strategies in the sequence from low-cost to high-cost one.

6.1.3.4 Levels of risk coping

According to the levels involved in the risk coping, all the strategies were classified into 4 categories: household, extended family, above extended family and formal insurance. As shown in Table 6-6, statistics demonstrate that the household is the basic unit of risk coping. Of 403 times of coping, 289 times (or 71.7%), took place within the household. All interviewed households stressed the imperative to make use of their own resources to solve their economic problems before requesting or receiving economic support from outside. It is only when intra-household coping proves insufficient that the extended family begins to operate. Of 403 times of coping, 87 (or 21.6%) are involved in this level. The transfers from relatives or good friends are especially important in the cases of catastrophic illness of a productive household member. In the entire sample households, people can deal with risk of illness well by the aid of the relatives, the cumulative percentage of the two levels amounts to 93.3 percent.

Only 6.7 percent of all coping strategies involve in above the level of extended family or formal health insurances. No sample household has got the loan from the rural financial institutions, although applying for a short-term consumptive loan became possible recently. The support from the community as a whole was also quite rare, but some neighbors helped a bit.

Table 6-6 Levels of coping strategies

Level	Household	Extended family	Above extended family	Health insurance
Coping strategy	Cash or savings, sale of livestock, sales of other assets (consumptive durables, cereals etc.)	Private transfer in the forms of gift, cash, and labor; Obligatory transfer; interest-free borrowing	Allowing for payment delay, free care, gift-transfer within local community, government relief, loan from financial institutions or moneylenders	Repayment of health insurance scheme, commercial insurance policy
Count	289	87	10	17
%	71.7	21.6	2.5	4.2

Source: ZEF survey 2002

6.1.3.5 Success of risk coping

As to the first criterion laid out above, i.e., calamity aversion, coping strategies were highly successful: only one of the households disintegrated or migrated as a consequence of illness costs. The second criterion is concerned with the degree to which coping succeeds in averting adverse effects on household production. Most households could maintain household production by substituting labors lost due to illness, but it did not matter even they could not deal with risk intra-household. They got labor transfers readily from their relatives. With respect to the third success criterion, whether coping behavior jeopardizes the household's asset base, the majority of households that purchased health care did not sell livestock due to illness.

6.1.3.6 Main determinants of coping strategy adoption

Until now it has been investigated how the rural households cope with financial costs of illness. But it is of further interests to unveil the factors underlying the coping strategies adoption. The main factors include income, medical costs, care provider, household size as well as some attributes of household head. Their relations revealed by the statistics are presented as follows.

Income and coping strategies

Statistics indicate that annual income per household determines which strategy the household adopted as the first step ($p=0.014$). To examine the effect of income on the coping strategies adoption, all the sick members are divided into three categories according to the level of total annual income of the household to which they belong: "poor", "average" and "rich". As seen in Table 6-7, some 96 percent of rich households settled health care bills of the sick member directly with their own cash or savings, while some 82 percent of the average and 75 percent of poor households adopted this method. The percentages of the poor and average categories are statistically significantly lower than that of the rich category.

Table 6-7 Income group vs. the main first coping strategy adoption

Type of coping strategy	Group of total income						Total	
	Poor		average		rich			
	Count	%	Count	%	Count	%	Count	%
Own cash or savings	49.0	75.4	70.0	82.4	98.0	96.1	217.0	86.1
Sale of livestock	7.0	10.8	4.0	4.7	0	0	11.0	4.4
Sale of other assets	1.0	1.5	1.0	1.2	1.0	1.0	3.0	1.2
Loan from relatives	3.0	4.6	6.0	7.1	2.0	2.0	11.0	4.4

Source: ZEF survey 2002
Note: income group is classified according to total annual income per household

As far as the last resort is concerned, it can be seen from the last column in Table 6-8 that "transfer from relatives" plays the most important role, followed by "reimbursement of insurer". It is also seen from the last row of Table 6-8 that the poor households are more likely to adopt a third round of strategies. Taking a close look at the last resort of different income groups, the rich households never used "sale of livestock", "sale of other assets", "loan from moneylender", and "payment delay" because the costs of these strategies are markedly high. For example, sales of livestock and assets may become distress sales, that is, they may be sold at lower prices than usual. Loans from moneylenders are not interest free. In contrast, the poor group never used "reimbursement of insurer" and "loans from relatives" whose costs are relatively low. However, the strategy of "transfer from relatives" seemed unbiased and was equally used among the three income groups. To sum up, the rich are more likely to be able to use low-cost strategies while the poor have to employ the high-cost strategies.

Table 6-8 Income group vs. last resort of risk coping

	Group of aggregate annual income			Total
	Poor	average	rich	
Sale of livestock	2	1	0	3
Sale of other assets	2	1	0	3
Loan from relatives	0	2	1	3
Transfer from relatives	9	6	4	19
Loan from moneylender	2	0	0	2
Allowing a lagged payment	1	0	0	1
Others 01(obligatory transfer)	0	0	1	1
Others 02 (reimbursement of insurer)	0	3	4	7
Total	16	13	10	39

Source: ZEF survey 2002

Medical expenditures, first contact provider vs. coping strategies

The selection of coping strategies is also significantly correlated to the magnitude of health care expenditures. The higher the user fee is, the more likely a household adopts the strategies of sale of poultry/livestock, other assets, loans from relatives, and transfer from relatives. Among the four user fee groups, furthermore, the rates of adoption of own cash or savings differ significantly, ranging from 96.5 to 76.2 percent as seen in Table 6-9.

It is expected that the large majority of people who seek care in a village health station should be able to cover the user fees by means of savings or cash and a small fraction of people who seek care in a county hospital be able to pay off the medical bill with cash and savings because the care therein is most costly. The statistics is partially consistent with this expectation. As seen in Table 6-10, despite the low costs of health care in village health stations, the percentage of using own savings and income to cover the costs of health care is the lowest (80%). A possible reason is that the people who seek care in village health stations may be especially poor and even the lowest user fees should be raised by other means.

Table 6-9 Quartiles of user fees vs. coping strategies

	Quartile of user fees								Total	
	1st		2nd		3rd		4th			
	Count	%	Count	%	Count	%	Count	%	Count	%
Own cash or savings	55	96.5	59	89.4	54	81.8	48	76.2	216	85.7
Sale of poultry/livestock	1	1.8	1	1.5	4	6.1	5	7.9	11	4.4
Sale of other assets	0	0	0	0	2	3.0	1	1.6	3	1.2
Loan from relatives	0	0	3	4.5	4	6.1	5	7.9	12	4.8

Table 6-10 The first contact provider vs. coping strategies

	Village health station		Town health center		County hospital		Private facilities		Total	
	Count	%	Count	%	Count	%	Count	%	Count	%
Own savings or income	39	79.6	85	91.4	38	79.2	53	86.9	215	85.7
Sale of poultry/livestock	5	10.2	2	2.2	3	6.3	1	1.6	11	4.4
Sale of other assets	0	0	2	2.2	0	0	1	1.6	3	1.2
Loan from relatives	3	6.1	2	2.2	3	6.3	4	6.6	12	4.8

Source: ZEF survey 2002

Household size and coping strategies

The household size versus the diversity of coping strategies is tabulated in Table 6-11. It demonstrates that the household size significantly affects the decision of coping strategies adoption. As for the first round of coping, the feature is that the choices of households of different size vary considerably. Households of 4 members have diversified their choices most, all the strategies on the menu are at the disposal of 4-member households; households of 3 or 5 members have the second most choices; both the small-sized households (of 1-2 members) and the large-sized (of 6-9memebrs) adopted few strategies similarly, which is not as diverse as the strategies of medium-sized household.

Concerning the 3rd round of coping, a similar regularity of coping strategy adoption can be observed. The medium-sized households (of 4, 3 and 5 members) have the most options while the small- and large-sized have fewer options. Despite the similarly few options of strategies of the small- and large-sized households, the reasons behind them may be different. The small-sized households have a small base of choices, whereas the large-sized have a large base of choices at their disposal but are not needed to take any other actions except using own cash or savings. It can be interpreted (1) that the households' wealth is positively correlated with the household size, the larger the household, the more the wealth as shown in the previous chapter; and (2) the medium-sized households have a stable household structure (husband, wife and kids) and are reliable for donors, compared with the single-member household or single-parent households.

Table 6-11 Diversity of coping strategies and household size

Household size	Number of 1st coping strategies	Number of the 3rd coping strategies
1-2	3	1
3	5	5
4	8	6
5	4	4
6	1	4
7-9	1	1

Source: ZEF survey 2002

Note: the second round of strategies are ignored and the third round is treated as the last resort.

Ethnicity and education of household head versus coping strategies

As shown in Table 6-12, Han people are quite likely to cope with costs of medical care with cash and savings. By contrast, minority people have to adopt other high-cost strategies. For example, some 18 percent of minority households and only one percent of Han households sold poultry/livestock to cover the user fees charged by first contact health facilities. The difference resulted from ethnic group, moreover, is statistically significant. It may be derived from two facets. One is due to the wide gap between the medical costs and household monetary income. The other is the extremely underdeveloped rural finance market in minority areas that makes monetary savings hardly possible.

It is expected that the household head's education is an important factor underlying the coping strategy adoption. The statistics indicate that the heads with lower education (illiterate, elementary school, junior high school) have similarly low likelihood of using cash and saving whereas the household heads with other education level have a significantly high probability of using cash and savings to cover the medical care. It is important to note that the probability of using own cash and savings is not proportionate to the level of education. For instance, the rate of college graduates using cash and savings is even lower than that of senior high school graduates. This may be attributable to a high likelihood of college graduates using reimbursement of insurance.

Table 6-12 Ethnicity, education of head vs. coping strategies

	Own savings or cash		Sale of poultry /livestock		Sale of other assets	
	Count	%	Count	%	Count	%
Ethnicity						
Han	185	91.1	2	1	3	1.5
Non-Han	31	63.3	9	18.4	0	0
Education of head						
Illiterate	45	84.9	0	0	2	3.8
Primary school	55	82.1	4	6	0	0
Junior high school	55	83.3	6	9.1	1	1.5
Senior high school	51	92.7	1	1.8	0	0
College	10	90.9	0	0	0	0

Source: ZEF survey 2002

County and coping strategies

Not amazingly, the coping strategies adoption varies across counties as shown in Table 6-13. On the one extreme, 100 percent of sample persons in Dongpo district used their own savings or cash to pay the medical care bill in the health facilities they contacted first. On the other extreme, people in Yanyuan were least likely to use this strategy (63.3%). Additionally, people in the county of Jiange were more likely to take some high-cost strategies such as the sale of poultry/livestock. The share of medical care costs in the household income may well interpret the phenomenon. As shown in Table 6-3, the low share of medical care costs in household income (5.1%) of Dongpo district allowed households to be able to cover the costs with own cash or savings. In contrast, the high share of medical care costs in household income (47.3%) caused the household in Jiange to use up low-cost strategies firstly, and then to rely on some high-cost strategies. Despite the low share of medical care costs of Yanyuan (16.5%) in comparison with Jiange, the people in Yanyuan relied more on high costs strategies because they have less monetary income and savings due to the exceedingly underdeveloped produce and financial markets, but the payment for medical care had to be made in cash.

Table 6-13 Coping strategies by Counties

	Jiange		Fushun		Yanyuan		Hanyuan		Dongpo	
	Count	%	Count	%	Count	%	Count	%	Count	%
Own savings or cash	22	88.0	33	89.2	31	63.3	50	82.0	80	100
Sale of poultry/livestock	1	4.0	0	0	9	18.4	1	1.6	0	0
Sale of other assets	1	4.0	1	2.7	0	0	1	1.6	0	0
Loan from relatives	0	0	1	2.7	8	16.3	3	4.9	0	0

Source: ZEF survey 2002

6.1.4 Discussion and policy implication

Direct costs of illness and disease-induced poverty

Although the actual costs of illness is always much greater than the amount of money that an individual or the society spends, economic costs remain a useful measurement of the magnitude of costs of illness. The previous research usually computes the direct economic costs of illness as the sum of user fees charged by a provider and the value of goods and services that is not produced because of illness. For example, Asfaw (2002) computed the monthly economic costs of illness per sick person in 15 sample villages by this means and average monthly direct costs of illness per sick person was US$13.64 or around 4 per cent of average annual income of household. The aggregate average monthly direct costs of illness including financial and time costs in Sichuan of China are lower than that of Ethiopia (US$9.29). The time costs of illness in Sichuan, China (US$4.19) are similar to that of Ethiopia (US$4.67) and much lower than that of Burkina Faso (US$8.27) in 1992 (Sauerborn *et al* 1996a). The proportion of yearly direct costs of illness in the annual household income in Sichuan may amount to an average of 29.0 percent, which is lower than that of Ethiopia (48%). Thus, average yearly direct costs of illness in Sichuan are moderate.

The computation has one additional important implication, that is, we can see how costs of illness are related to poverty. In different places of China, illness has induced 20 to 70 percent of poverty as summarized in section 5.1. This research shows that the illnesses in Jiange, Hanyuan and Yanyuan are quite likely to induce poverty because of their high proportions of costs of illness in household's income. In particular, average user fees per household in Jiange accounts for 47.5 percent of household income; this has affected the daily consumption of the household because average Engle coefficient in rural areas of this province is about 0.55 (NSB 2002). Obviously, the user fees may exceed the disposable income of household easily; furthermore, the time of earning monetary income may not necessarily coincide with that of payment of user fees. Therefore, this time difference necessitates the adoption of high-cost strategies, such as sales of productive livestock, impairing the capacity of earning income in the future and causing the household to live in poverty.

The role of the household and extended family in coping

As many previous studies observed, the basic unit of risk coping is the household rather than individual (e.g., Sauerborn *et al* 1996a and literature cited therein; Dercon and Krishnan 1996). Our statistics indicate that 72 percent of health risks can be coped with successfully within household. Not surprisingly, the household characteristics such as income, ethnicity, education and age of household head are more related to the coping strategies adoption than those of individuals. The most important determinant is the proportion of medical care costs in annual household income. But even within the household, the possible strategies are different from each other: cash and monetary savings can be used with almost no cost, while the sales of livestock or assets are of moderate cost. The household is important in risk coping partially because it provides a mechanism for other strategies to work. For example, the loan provider would run a big risk of default if the loan were to be repaid by the sick her- or himself, particularly when the sick is old.

The extended family that is composed of relatives' and good friends' households also plays an imporatnt role in risk coping. Some 22 percent of coping strategies are related to relatives and good friends. It is noteworthy that the role of extended family must be underestimated if we judge it simply from the percentage. Severe health events are infrequent after all, but the assistance from other households is usually involved once severe health events happen. Its important role in risk coping is to be concretely and completely addressed in the next section.

6.2 Role of the extended family in coping with health risk

Perhaps the most universal institution of pooling resource and sharing risk is the household (Dercon and Krishnan 2000). One of the major motives behind household formation is to provide care and nurturing for the most defenseless among us: children. In rural communities, households are also the institution within which care is taken of the sick, the disabled and the old. It is within the household that people find solace when they are hit by disease and that they seek moral support when they are struck by bad luck. The household's role can continue beyond the household disintegration, many of the legal and customary rules governing the devolution of assets upon the household disintegration can be seen as an effort to protect its most vulnerable members, children and, to a less extent, wives. Given the important role of the household in sharing risk, it is not amazing to observe that large-sized households in some high-risk environments have five to eight members on average and comprises several nuclear households. The structure and organization of these large households can be quite complex and each member or nuclear unit is often granted quite a bit of autonomy in managing their own affairs provided they fulfill their obligations to the common goods of the household (e.g., Binswanger and McIntire 1987; von Braun and Webb 1989). From the opposite perspective, the important role of the household in sharing risk can also be seen *in absentia* by the fact that the poorest members of society are usually isolated members: the childless aged with nobody taking care of them, runaway children, physically and mentally disabled individuals who have been abandoned by their family.

The household may become small if the risk the household faces decreases, because the benefit from the risk pooling on the condition of the lowered risk may be offset or outweighed by high managing and monitoring cost of large households, especially those with many nuclear units. Naturally these large households are split into many small households including only one nuclear unit in each while all the nuclear units reside not far and keep close connections. This is known as the *extended family (EF)*. It is defined as a loose and informal organization that is composed of households of relatives and good friends. Although the households in the extended family are still sharing risk, the form of risk sharing differs. The large household shares risk among members by united organization and management. But how does the extended family function as a risk sharing arrangement? The question is to be answered theoretically in the following section.

6.2.1 Theory on the extended family as a risk sharing arrangement

6.2.1.1 An infinite repeated game model

Compared with the coping strategies within a household like income diversification and precautionary savings, risk sharing (whether implicit or explicit sharing) across many households is able to cope with not only big idiosyncratic risks but also some of covariate risks. Given that rural people tend to be risk averse while living in a high-risk environment, there should have been many mutual insurance arrangements in society to cope with these risks. In reality this is not the case. A game model may be conducive to understand this point. Let's start with a finite repeated game model simulating general informal risk sharing in a rural community (Fafchamps 1999) and gradually adapt it to an infinite repeated game model in the context of the extended family.

General informal risk sharing in a rural community

Let s_t denote the state of the world at time t and be drawn from a finite set of possible events S_t with probability Prob(s_t). Consider a rural community constituted of N members, each with a stream of stochastic income $\pi_{it}(s_t)$ and a utility $U_i[c_{it}(s_t)]$. Pareto efficiency in the sharing of risk requires that for any set of welfare weights of ω_i individual consumption satisfies the following social planner problem.

Eq. 6-1 $$\max_{\{c_{it}\}} \sum_{i=1}^{N} \omega_i \sum_{t=1}^{\infty} (1+\delta)^{-t} \sum_{s_t \in St} \Pr ob(s_t) U_i(c_{it}(s_t))$$

subject to the feasibility constraint (namely aggregate consumption equals aggregate income)

Eq. 6-2 $$C_t(s_t) \equiv \sum_{i=1}^{N} c_{it}(s_t) = \sum_{i=1}^{N} \pi_{it}(s_t) \equiv \prod_t (s_t)$$

and a set of non-negativity consumption constraints

Eq. 6-3 $c_{it}(s_t) \geq 0$

for all s_t and all t. The set of all efficient risk sharing allocations can be traced by varying the welfare weights ω_i.

This social planner model can easily be extended to allow for accumulation. Given the existence of a riskless asset A with constant return r, the feasibility constraint becomes:

Eq. 6-4 $$C_t(s_t) \equiv \sum_{i=1}^{N} c_{it}(s_t) = \sum_{i=1}^{N} (\pi_{it}(s_t) + A_{it}) \equiv X_t(s_t)$$

where $X_t(s_t)$ denotes aggregate cash in hand.

Let $\lambda(s_t)$ be the Lagrange multiplier associated with the s_t feasibility constraint. Assuming that none of zero consumption constraints are binding, efficient risk sharing requires that:

Eq. 6-5 $$\omega_i (1+\delta)^{-t} \Pr ob(s_t) U_i^{'}(c_{it}(s_t)) = \lambda(s_t)$$

Since $\lambda(s_t)$ depends only on aggregate income $\Pi_t(s_t)$ (with no accumulation) or aggregate cash in hand $X_t(s_t)$ with accumulation, Eq. 6-4 implies that Pareto efficient allocation individual consumption can only depends on $\Pi_t(s_t)$ and $X_t(s_t)$, not on individual income.

The Pareto efficiency conditions for an interior solution can be rewritten:

Eq. 6-6 $$\frac{U_i^{'}(c_{it}(s_t))}{U_j^{'}(c_{jt}(s_t))} = \frac{U_i^{'}(c_{it}(s'_t))}{U_j^{'}(c_{jt}(s'_t))} = \frac{\omega_j}{\omega_i}$$

From Eq. 6-6, it is easy to see that if the welfare weights and utility functions are the same for two agents an i and j, then:

$$c_{it}(s_t) = c_{jt}(s_t)$$

for all t and s_t. In this case, risk sharing is equivalent to pooling incomes and sharing them equally.

Risk sharing, self-interest and commitment failure

Risk sharing in a rural community is not devoid of difficulties. It is expected to be close to an infinite repeated game. However, the rapid social movement in rural community, e.g., migration, usually makes game play a finite one, which in essence is not different from one-shot game. The factor underlying commitment failure is self-interest. To see why, note that mutual insurance is Pareto improving whenever agents are risk averse and risk is at least partly idiosyncratic. Consequently, there exist informal mutual insurance contracts to which agents would voluntarily agree *ex ante*. In other words, if $U_i(.)$ is concave and $\pi_{it}(s_t)$ varies across individuals, then there exist a set of contingent promises $\tau_i(s_t)$ such that

Eq. 6-7 $$EU_i(\pi_{it}(s_t)-b_{it}(s_t)+\tau_i(s_t))\geq EU_i(\pi_{it}(st)-b_{it}(s_t))$$ for all i

with $\sum_i \tau_i(s_t)=0$ for all t and s_t.

where b_{it} is an individual-specific shifter that captures household characteristics and utility shocks such as disease or rituals. According to the prediction of one-shot game and finite repeated game models, 'never assisting others' is the Nash equilibrium and the sub-game perfect Nash equilibrium. Although agreeing the informal contract $\tau_i(s_t)$ is always individual optimal *ex ante*, yet a rational and self-interested agent will never respect the contract whenever $\tau_i(s_t)<0$, that is, when they are asked to assist others. Of course, if nobody ever assists others *ex post*, a commitment failure occurs: mutual insurance is not achieved and the outcome is sub-optimal. A question arises: what mechanisms can be used to ensure that the commitment failure is avoided and Pareto gains are achieved? The most often-encountered types of system, according to Fafchamps (1998), are social insurance, family values and long-term relationship. Owing to the uptight governmental budget, the developing countries cannot afford the universal social insurance, including rural areas. The researcher will provide a solution combining family values with long-term relationships—the extended family-based risk sharing mechanism.

The extended family and reciprocal private transfers

One way to minimize the commitment failure other than resort to social institutions or the legal system is for individuals to enter into long-term relationships with each other. The inspiration is derived from the solution to prisoner's dilemma that an infinitely repeated game could help prisoners out, according to the folk theorem[26]. Kimball (1990) and Coate *et al* (1993) formalized the application of game theory to informal risk sharing arrangements by demonstrating that an implicit risk-sharing agreement can be sustained through repeated interaction and thus that commitment to assisting others can be self-enforcing. Consider an exchange economy in which output cannot be stored, assets cannot be accumulated, and there is no borrowing and lending from the rest of the world. People derive utility from what they consume $U_i(c_s^i)$, and at least some people are risk averse, i.e., $U''_i < 0$ for some i's. Let $\delta \in (0,1)$ be a common discount factor and let Q denote a sequence of action profiles or path of the economy. Define $\omega_i(Q)$ as agent i's discounted expected payoff along that path, i.e.:

[26] Even in finite repeated game, its sub-game perfect Nash equilibrium remains the same as the Nash equilibrium in one-shot game. See Fudenberg, Dew and Jean Tirole, game theory p152-153, 1991, the MIT press.

Eq. 6-8 $$\omega_i(Q) \equiv \sum_{t=0}^{\infty} \delta^t EU_i(\pi^i_{s,t} - \tau^i_{s,t}(Q))$$

where the $\tau^i_{s,t}(Q)$ refers to actual transfer at the end of each period t dictated by action profiles Q.

Following Abreu (1988), $N+1$ strategy profiles are sufficient to span the complete set of equilibrium payoffs of this economy: one cooperative risk sharing strategy profile denoted Q^0, and one punishment strategy profile Q^k for each of the $k \in \{1,\dots, N\}$ agents. Agents play according to Q^0 as long as nobody deviates, and switch to Q^i following a defection by agent i either to the initial path Q^0 or to any of punishment Q^k. Abreu (1988) showed that adding other more complex strategies cannot expand the set of equilibrium payoffs—and thus the extent of risk sharing, provided that punishments Q^k are the harshest that is able to credibly inflict on the deviant agents. Voluntary participation in informal risk sharing implies that agents cannot be maintained below the expected payoff they could guarantee themselves by exiting the risk-sharing group. Since autarky is the Nash equilibrium of the one-shot game in rural community, autarky payoffs serve as the harshest punishments:

Eq. 6-9 $$\omega_k(Q^k) = \frac{EU_k(\pi^k_s)}{1-\delta} \quad \text{for } k \in \{1, \dots, N\}$$

Along the equilibrium of path Q^0, net transfer between agents depends not only on their own realized income but also on that of all other agents. This imposes high information requirements since agents must monitor each other's income to spot defections (e.g., fafchamps 1992; Ligon 1996). Rural communities in which households make a living mainly through agricultural production that is easily observed is expected to have more informal risk sharing arrangements.

The set of sub-game perfect Nash equilibriums of an infinite repeated game is large and the payoffs are comprised between the Pareto efficient frontier and a set of voluntary participation (VP) constraints that must be satisfied along any equilibrium path after the realization of s:

Eq. 6-10 $$U_i(\pi^i_{s',t}) - U_i(\pi^i_{s,t} - \tau^i_{s',t}) \le \delta\omega_i(Q^k) - \delta\omega_i(Q^i) + A_i$$

for all $s' \in S$ and all $k = 0, 1, \dots, N$. Participation constraints require that for any realization of the state of nature s', the short-run gain from deviation $U_i(\pi^i_{s',t}) - U_i(\pi^i_{s,t} - \tau^i_{s',t})$ be smaller than the discounted long-run gain from cooperation $\delta\omega_i(Q^k) - \delta\omega_i(Q^i)$. It is clear from the above equation that VP constraints are never binding when $\tau^i_{s',t}$ <=0. Voluntary participation is problematic only when Q^k requires an agent to help others, when $\tau^i_{s',t}$>=0. A_i (≥ 0) captures the subjective satisfaction that agents may derive from 'doing the right thing'. Alternatively, A_i can represent the guilt that people may feel for reneging on their promises. Social sanctions other than exclusion from the risk-sharing group are included in A_i.

Nevertheless, how the long relationship should be maintained to sustain the arrangement remains unanswered. From the right hand of Eq. 6-10 we can see that the informal risk-sharing arrangement will be more likely to be sustained if three conditions are satisfied: (i) high discount factor, namely, players are very patient; (ii) harsh punishment, namely, the autarky is quite undesirable; (iii) high Ai, that is, the subjective satisfaction from doing the right thing or the guilt from reneging on promises is very large. From the left hand of the equation we find small $\tau^i_{s',t}$ relative to average income is a good solution.

According to these conditions, the researcher argues that the *frequent and small-scale reciprocal private transfers within the extended family* can sustain informal risk-sharing mechanisms well. As stated previously, both the legal and moral rules stress the role of family in coping with risk and hence family value A_i will acquire continuous strengthening. For instance, one would feel much more guilty when failing to assist his/her brothers than when failing to assist an ordinary fellow villager. In the mean time, one may incur more moral condemnation and social sanction. Reciprocal private transfers may also be signs of faithfulness. Others may feel one is not trustworthy if he unexpectedly does not assist his blood relations. How can ordinary people count on him? Member households within the extended family still need to show or renew their close relationships through regular reciprocal transfers even when no risk events take place. Thus, it is not difficult to understand the frequent reciprocal transfers between the member households within the extended family, whether they are made to assist a person in need of help or to demonstrate faithfulness. Formally this means that we temporarily restrict our attention to stationary strategies, that is, to strategies that depend only on the current state of nature s, not the past transfers. Strategies are affected by the past history of play only inasmuch as defection and punishments are concerned. In this case, VP constraints take a simpler form:

Eq. 6-11
$$U_i(\pi^i_{s'}) - U_i(\pi^i_s - \tau^i_{s'}) \leq \frac{\delta}{1-\delta} E[U_i(\pi^i_s - \tau^i_s) - U_i(\pi^i_s)] + A_i$$

In the equation the discount factor remains the key to sustain informal risk-sharing arrangement, high δ is desirable. As δ decreases, gains from risk sharing must be shared more equally. Thus, the small-scale and frequent reciprocal transfers within the extended family can maintain the informal risk sharing arrangements. Next, we try to find some empirical evidences in support of risk sharing within the extended family and to show that the *frequent reciprocal transfers within the extended family* under a particular socio-economic and cultural context may have a more stable and larger set of sub-game perfect Nash equilibriums than those under other contexts.

6.2.1.2 Evidence of risk sharing within extended family from literatures

In the last subsection instability of sub-game perfect equilibriums among households in a large rural community have been theoretically demonstrated.

The equilibrium may readily suffer commitment failure, which may cause the informal insurance arrangement to break down. It was also illustrated how the set of sub-game perfect equilibriums will expand by strengthening *family value (Ai)* and *long relationships*. The discount factor does not need to be that large in order for the players to be patient enough. Let's see some evidences from empirical researches.

It has been widely tested that rural households at the level of community are not completely insured by some informal insurance arrangements (e.g., Dercon and Krishnan 2000; Townsend 1994; Morduch 1991). However, according to Morduch (1991), efficient risk sharing cannot be ruled out within sub-groups such as caste, or for food consumption. Some economists suggest that the extended family (relatives plus friends) should be a potentially suitable risk-sharing group for testing the existence of complete insurance. Hoogeveen (2003) tried to test the presence of full insurance at the community level in rural Zimbabwe while taking into account the effect of an accumulation buffer stock. Yet even with assets included in the estimation, evidence in support of full insurance could not be obtained, neither did the evidence of absence of informal insurance arrangements. That is, rural households are only partially insured. Another suggestion made in his study is that had it been possible to define the relevant insurance group as the village plus *relatives* living elsewhere, full insurance might not have been rejected.

Gertler and Gruber (1997) take a different cut on test of risk sharing. In evaluating the ability to cope with the costs of illness in Indonesia, they find households insure adequately against about 70 percent of common health shocks but can protect their consumption levels against 30 percent of the illness that seriously impairs long-term performance. A study using data of rural Philippines (Lund and Fafchamps 1997) find further and interesting results in coping with costs of illness. Informal insurance arrangements are effective only in the case of young adults who are acutely ill; older adults who fall ill are far less likely to be helped. Lund and Fafchamp's data show that informal insurance also helps with funerals but not with crop failures, mild illnesses, or unemployment. The household's social network also matters: households with more friends (especially richer friends) have a greater ability to use informal insurance. Households that are not so well connected fare far worse.

Normally the extended families operate through reciprocal private transfers in the forms of gift and money. When risk events such as sickness take place in one member household, other members will transfer gift or money to the ill just like an insurance company to compensate the insured for their financial loss. Labor is another common form of reciprocal transfers, that is, relatives and friends help attend the sick or even work in the field. The magnitude and frequency of private transfer vary across countries. Cox *et al* (1995) found that 82% of the urban and 89% of the rural households in the Philippines reported receiving transfers regardless of reasons. Sauerborn *et al* (1996a) examined the strategies that the rural households in Burkina Faso used to cope with the costs of illness and found that despite many self-insurance strategies such as deposit drawing and sales of assets, 14 of 25 households with illness events received gifts from extended kin. However, the purpose of gift giving is not necessarily to deal with risk events as we argued before. It may also be used as a sign of commitment to the group. So it is not surprising to see that in North China gifts are given to mark births, deaths, and weddings, besides to help the elderly, the ill, and women who have just given birth (Kipnis 1997).

The importance of the reciprocal transfers within the extended family may be confirmed from the reverse—transfers from outside of the extended family are relatively small and infrequent. Morduch and Sicular (1999) found that no more than 25 percent of households in North China reported receipts of transfers from their neighbors; 10 percent reported receipts of gifts from outside the village. The magnitude of transfers is modest, about 10-20 percent of average household income.

6.2.1.3 Informal and formal insurance arrangements

Until now, we mainly focused on the informal insurance arrangements. Considering the similar functions between informal and formal insurance arrangements (public social insurance, community-based health insurance scheme and private health insurance), a natural question arises: Is there conflict between the formal and informal insurance arrangements? The conflict between the informal insurance arrangements and public social insurance has been implicitly assumed and confirmed. However, attention is only paid to the direction from the formal to the informal. For example, Cox and Jimenez (1995) find that those who are most likely to receive private transfers from relatives and friends would also be eligible to receive assistance through public safety net programs. This can create a problem known as *crowding out* of the private transfers by public programs (Cox and Jimenez 1990; Lampman and Smeeding 1983). Crowding out is a cause for policy concern because it implies that public resources are inefficiently allocated. The creation of crowding out is to a large extent owing to the direct intervention by the state and it has drawn much attention from researchers and policy makers.

With the development of the market system in developing countries, the community health care financing schemes and for-profit commercial insurance enterprises play an increasing role in risk management of rural households. However, there has been little attention paid to the possible conflict between the informal and formal arrangements. This deserves further investigation.

Unlike the public insurance programs implemented by the governments, community financing schemes and commercial insurances do not have privileges to force the rural households to participate in the schemes and purchase policies. The stable and normally functioning informal insurance arrangements may hinder the voluntary mechanisms from crowding in.

This section shall focus on seeking empirical evidences on the risk sharing at the level of extended family and on testing the hypothesis that the effect of crowding out also exists between the extended family versus rural cooperative medical scheme. Let's turn to the setting of rural China, as predicted by the infinite repeated game model. A well functioning informal insurance arrangement should meet three conditions: (i) high discount factor, namely, players are very patient; (ii) harsh punishment, namely, the autarky is quite undesirable; (iii) high Ai, that is, the subjective satisfaction from 'doing the right thing' or the guilt from 'reneging on promises' is very large. The extended family in the setting of Chinese culture basically satisfies these conditions, the reasons are enumerated as follows.

First of all, deeply influenced by the Confucian culture for more than 5000 years, social networks are quite common in Chinese society and play an important role in many respects (Jiang *et al* 2003 and literatures cited therein); Chinese also assign a high value to family and put much emphasis on kinship. For example, people usually believe that 'blood is thicker than water'. The promises of private transfers between relatives may even carry over generations. Thus, the condition of 'being patient' and 'high A_i' of the infinite repeated game should be easily satisfied. Next, since the Communist Party of China (CPC) founded the new government in 1949, the policy of a high fertility rate pursued by the government for three decades has led to the common existence of large extended families. In contrast, the policy of family planning, 'one couple one child', implemented in the recent two decades to reverse the rapid population growth made small households composed of only one nuclear unit popular. With only one major male laborer and with no assistance of other households, one rural household is unable to undertake many heavy items of agricultural production and other income-generation activities because the farm activities are usually under-mechanized. Thus, autarky is quite undesirable. Finally, the absence of social security such as rural pension insurance and health insurances offers a tremendous space for the development of informal insurance arrangements.

6.2.2 Methodology and descriptive analysis

6.2.2.1 Methodology

On the theoretical and empirical bases, a basic assumption in this section is that risk sharing within the extended family has already existed. We then try to investigate how the extended family works as a risk sharing arrangement. Inspired by the mechanism of formal insurance arrangement, we try to measure its size (equivalent to risk pool in insurance company), composition, frequency of transfers (probability of risk), magnitude of transfers (scale of compensation) and geographic barrier. We are also concerned about the interaction between the extended family and formal insurance participation. A testable implication is that whether the extended family negatively affects demand for health insurance significantly (in terms of observed demand or potential demand).

For this purpose, we designed one section in our questionnaire to obtain overall information on the reciprocal transfers within the extended family. According to the previous statement, one year may be too short to meet the condition of an infinite repeated game. Thus, we select the period of time "since you had memory and to your best knowledge". Although the recall may be not quite accurate, yet for some catastrophic risk events, the household head or the spouse should be able to recall what and how much transfers had received from relations and extended kin or good friends (e.g., one member became disabled in an accident).

Statistically, two types of nuclear households are counted in the extended family. First, all households with direct blood-relations with the respondent's household: parents, brothers, sisters, sons, daughters, and step children or foster children's of the household heads or their spouses, regardless of whether they have made a transfer; second, collateral relatives and good friends who have made transfer to the respondents' household at least once. Aunts, uncles, cousins, nephews, nieces, some scotch cousins and good friends fall into this type.

The transfers are divided into two types: transfers due to illness and transfers for other reasons, which are explicitly indicated in the questionnaire, including pension, wedding, housing construction, funerals and so on. Transfers are recorded in two forms: monetary values if transfers are in the forms of gift and cash, and days if transfers take the form of labor. Our main attention is focused on the former. All the gifts are converted into monetary value in this way: the enumerators record the gift names first, and then ask the respondents to make an estimation over the monetary value of the gifts.

6.2.2.2 Descriptive analysis of private transfers within the EF

The size and composition of the extended family

Statistics shows that all the 300 sampled households have received private transfers because of illness of household members from at least one other household that is either a relative or a good friend. Some of households receive transfers from 21 other households. On average, a household has 9 households of close relatives, extended kin or good friends. According to the definition of the extended family (EF), the size of the extended family is an average of nine, excluding the respondent's household. Apart from the size, the composition of the extended family is of concern because the composition of the extended family reflects who make transfer and who are quite likely to make transfers. As shown in Table 6-14, 'brother' and 'sister' are the most reliable aid providers. Brother accounts for 32.6 percent of all the relatives and friends within the extended family, ranking the first, followed by sisters (26.2%). The sum of the two types accounts for 58.8 percent of the extended family. Parents and offspring are not common aid providers within the extended family if judged from our statistic results, only 0.9 percent and 2.4 percent of relatives and friends are sons and daughters, respectively and 6.6 percent are parents. This is, however, contradictory to the common sense. The explanation is that old parents usually resided with one of their sons who became the household head. But we only questioned the relationship between household head and relatives' household head. Thus, even though the transfer had been made from sons or daughters to their old parents, it was still recorded as "from household of a brother (or sister) to household of another brother" rather than "from son (or daughter) to parents" in the survey. Unfortunately, our data is unable to differentiate the former from the latter. In total, the direct relatives account for 70.5 percent of all relatives and friends. It is essential to note that of all the direct relatives, only 82 percent have once helped the respondents' households. Even so, they are still quite likely to help the respondents' households in the future considering the very close kinship.

Where the extended kin is concerned, the good friends and cousins are striking. Good friends are reliable donors for the respondents' households. Good friends and the cousins accounted for about 8 percent, 8 percent of the extended family, respectively. Comparing the aunts/uncles with the nephews/nieces, we find that the elder kin are slightly more important than the younger kin. Other scotch cousin accounts for only 1.1 percent. To sum up, the direct relations and near relatives plays a far more important role in risk sharing and the extended kin is still an integral component of the extended family.

Table 6-14 Composition of extended family by relation to household head

		Frequency	Percent	Valid Percent	Cumulative Percent
Valid	Parents	179	6.6	7.0	7.0
	Brother	840	31.1	32.6	39.6
	Sister	707	26.2	27.5	67.1
	Son	25	0.9	1.0	68.1
	Daughter	64	2.4	2.5	70.5
	Other close kin	12	0.4	0.5	71.0
	Cousin	218	8.1	8.5	79.5
	Aunt or uncle	137	5.1	5.3	84.8
	Niece or nephew	132	4.9	5.1	89.9
	Friends	228	8.4	8.9	98.8
	Other Scotch cousin	31	1.1	1.2	100.0
	Total	2573	95.3	100.0	
Missing system		128	4.7		
Total		2701	100.0		

Source: ZEF survey 2002

Demographic characteristics of the relatives within the extended family

As shown in Table 6-15, some 60 percent of the relatives within the extended family who are directly connected with the respondents' households are males and the remaining are females. Classified according to the location of the relatives' home, 84 percent are located in rural areas while 16 percent are located in urban areas. According to occupation of relatives, some 72 percent of relatives earn a living by farm labor, 7.7 percent move to the cities to do a casual job and 6.9 percent are employees in formal sectors insured by social insurance. Not all households in the risk-sharing group are uninsured, so being uninsured is not a necessary condition for joining the extended family risk sharing mechanism. The average age of relatives is 42.6 years of age.

Table 6-15 Demographic characteristics of relatives and good friends

Type		Frequency	Valid percent
Gender	Male	1539	60.2
	Female	1019	39.8
	Total	2558	100
Location	Rural	1793	84.1
	Urban	340	15.9
	Total	2133	100.0
Occupation	Farmer	1522	72.1
	Worker in VTE	66	3.1
	Village teacher	46	2.2
	Migrant worker	162	7.7
	Servants	87	4.1
	Urban employee	146	6.9
	Others	82	3.9
	Total	2111	100.0

Source: ZEF survey 2002

The wealth of the relatives and travel time

In order to examine the economic status of the relatives' households, the relatives are classified into three types according to their economic status compared with the respondents' households: richer, equal and poorer. The richer relatives accounted for some 39 percent, while the poorer relatives only accounted for some 17 percent. In other words, some 83 percent of relatives who have made transfer or have not made transfer but are most likely to do so in the future are at least not poorer than household questioned. The travel time from the respondents' homes to their relatives' homes is on average 1.3 hour. Some 79 percent of the relatives' homes are reachable within one hour by foot or by public transportation means. However, there are still 21 percent of relatives that live far away from the respondents' households. However, they may make transfer in the form of remittance, which is hardly affected by the geographic distance.

The frequency and magnitude of transfers with individual relative

Reciprocal transfer is a common way to solidify social and economic relationships. Not only the magnitude of transfer but also the frequency of transfer matter. Here, private transfers due to illness is distinguished from private transfers for other reasons to examine the role of the extended family in coping with the health risk. Results exhibit that, of 300 households, 284 (or 94.7 %) received private transfers in the forms of gift and cash at least once since the foundation of the household, while 214 (or 71.1%) made transfers to other relatives at least once.

Let's first have a look at the general situation of gift and cash receipts from the relatives. As shown in Table 6-16, the respondents' households received an average of 2.4 times of gifts and cash from individual relative due to illness. As for labor transfers, the average times of labor receipt are merely 3, but the transfers of labor are far more frequent than the figure reflects. Many respondents said that the labor exchange is so frequent that it is difficult to recall an exact number (as shown in Table 6-16, the missing system of labor transfer is very large). Labors exchange took place not only because of risk events, but also because of the consideration of economy to scale. For each of the small households that usually possess one major male laborer only, it is quite necessary to assist each other in undertaking heavy works that is undercapitalized in China, e.g., lumbering, reaping and home construction.

With respect to the magnitude of individual relative's transfer, on average the respondents received 315.6 Yuan of monetary transfer and 6.5 days of labor transfers from individual relatives due to illness. Seventy percent of single cash transfers from individual relatives are not more than 200 Yuan. Besides the magnitude of transfers due to illness, we also surveyed two times of biggest individual cash receipts to see how the sample households coped with the large costs of illness in the past. One of the biggest individual cash receipt averages 149.7 Yuan and the other 122. These have surpassed the mean reimbursement (121 Yuan) by rural cooperative medical scheme (RCMS) in Dongpo district. The biggest individual cash transfer happened around 6 years ago, that is, in 1996. The individual cash transfers for other reasons such as weddings, funerals, house constructions, pensions are much larger than that due to illness, averaging 982.9 Yuan, which are nearly 7 times as much as that due to illness. The low coverage of RCMS relative to the sum of monetary transfers makes RCMS unattractive for quite a few people.

Next, we turn to the general situation of gift and cash giving to the relatives. In the aggregate, the respondents' households gave much fewer gifts and cash to their relatives (623 times) by frequency due to illness than they received (1636 times). The average magnitude of individual gifts and cash giving to the relatives is 329.3 Yuan, a bit higher than the average gifts or cash receipts from the relatives. One of the biggest single donations due to illness is on average 173.1 Yuan, which took place in around 1997. The gift and cash giving to the relatives for other reasons are on average 699.6 Yuan, much lower than the gift and cash receipts from the relatives.

Table 6-16 Descriptive statistics on private transfer by households surveyed

	No of relatives	Mini.	Maxi.	Mean	Std. Deviation
Receipt					
Frequency of gift & cash receipt due to illness	1636	0	50	2.4	2.9
Frequency of labor receipt due to illness	550	0	50	3.0	4.3
GC receipt due to illness (Y)	1545	0	10000	315.6	710.7
Labor receipt due to illness (d)	471	0	200	6.5	15.4
One of the biggest GC receipt (Y)	1434	0	10000	149.7	413.5
Another of biggest GC receipt (Y)	676	8	4000	122.0	276.6
Year of one of the biggest receipts	1443	1968	2002	1996.0	5.4
Year of another of the biggest receipts	666	1975	2002	1998.6	4.2
GC receipt for other reasons (Y)	1791	1	35000	982.9	1868.6
Giving					
Frequency of GC donation due to illness	623	0	50	2.4	3.2
Frequency of labor donation due to illness	244	0	60	5.2	6.2
GC donation due to illness (Y)	612	1	5000	329.3	540.5
Labor donation due to illness (d)	248	0	200	9.7	20.9
One of the biggest GC donation (Y)	601	1	3000	173.1	321.3
Another of the biggest GC donation (Y)	253	5	2000	136.1	223.8
Year of the biggest donation	599	1980	2002	1997.1	3.9
Year of 2nd biggest donation	248	1983	2001	1998.8	2.9
GC donation for other reasons (Y)	697	0	7000	699.6	1006.4

Source: ZEF survey 2002

Note: GC is abbreviation of 'gift and cash'.

The Frequency and magnitude of aggregate household receipt and giving

Besides the frequency and magnitude of respondents' transfers with individual relatives, it is of more interest to see how many times and how much the respondents' households in total received from and gave to all the relatives and friends within the extended family. Results indicate that, since the foundation of the home, each respondent's household on average received 16 times of money or gift from its relatives and friends by frequency, and all the gift and cash receipts are worth 1627.1 Yuan. Although the figure is large, it is still mild if distributed among many years. The aggregate gift and cash receipts for other reasons is 3.6 times as many as the aggregate gift and cash receipts due to illness. Compared with the receipts, the donations of the respondents' households are much fewer. One household has on average donated gift or cash 7.1 times by frequency, which is in total worth 669.6 Yuan. The aggregate donations for other reasons are far fewer than that due to illness. In other words, the respondents' households tend to be net recipients.

Net gift and cash transfers and its relation with other factors

An attempt is made to compute the net gift and cash transfers by subtracting the gift and cash giving from the gift and cash receipt. The result is tabulated in Table 6-18. It shows that some 63 percent of the sample households are net recipients due to illness; some 74 percent are net recipients for other reasons such as weddings, funerals and house constructions. The proportion of net recipients (63%) due to illness is much higher than that of net donors (27%). The reasons might be (i) that our sample households are confined into rural areas and their main occupation is farmer, but their relatives could be living anywhere and on any occupation. Consequently, the rich and the urban relatives should be more likely to be net donors; (ii) that the respondents probably recall the receipts easily because each receipt is related to an impressive event at their own homes.

Table 6-17 Description of indicators measuring the risk-sharing function of the EF

	Number of households	Mini.	Maxi.	Mean	Std. Deviation
Relative size	300	1.0	21.0	9.0	4.0
Average age of relatives	296	27.2	70.0	42.6	7.2
Average travel time to relatives' home (Hr)	283	0.02	14.5	1.3	1.8
Receipt of gift &cash					
Times of receipt due to illness	247	0.0	193.0	16.1	19.7
Value of receipt due to illness (Yuan)	300	0.0	39360.0	1627.1	3447.2
Value of receipt for other reasons (Yuan)	299	0.0	75400.0	5887.7	9948.7
Giving of gift & cash					
Times of giving due to illness	214	0.0	140.0	7.1	11.9
Value of giving due to illness (Yuan)	300	0.0	15800.0	669.6	1455.9
Value of giving for other reasons (Yuan)	300	0.0	17400.0	1625.4	2739.9

Table 6-18 Net receipts of respondents' household

	Frequency	Percent	Cumulative Percent
Net GC receipt due to illness			
Net donors	81	27	27
Zero net transfer	31	10.33	37.33
Net recipients	188	62.67	100
Net GC receipt for other reasons			
Net donors	57	19	19
Zero net transfer	22	7.33	26.33
Net recipients	221	73.67	100

Source: ZEF survey 2002

Note: GC is abbreviation of 'gift and cash'.

Besides the number of net donors and recipients, it is also interesting what factors are related to being net donors or recipients. This can be statistically analyzed via a *contingent table* (Table 6-19). Results indicate that gender of household head appears to be related to being net recipient. Female household heads are more likely to be net recipients, regardless of the reasons of transfer. This consists with the research results from the USA, Philippine and Kyrgystan. In the USA, the probability of receiving private transfers is higher for female-headed households, young females, and for households with a large number of children (Moon 1983). In the Philippines, married females also have a higher probability of receiving private transfers (Cox and Jimenez 1995). The same results hold for Kyrgyzstan (Cox and Jimenez *et al* 1994). Nonetheless, the difference in this study is not statistically significant.

The elder-headed households are significantly more likely to be net recipients due to illness. But where net transfers for other reasons are concerned, the elder-headed households are slightly more likely to be net recipients. Research results from literatures on this point are diversified. Some studies found that private transfers usually flowed from the older to younger households (Schoeni 1993). In the USA, Schoeni (1993) found that unlike money transfers, labor assistance is also given to older households. Also, Ravallion and Dearden (1988) found evidence of significant transfers from the young to the old in Malaysia and Java. In our view, the transfers tend to flow from the young to the old in rural areas and from the old to the young in urban areas in China. The reason is that the young are compulsory to support for the elder who have no pension and health insurance in rural areas. This may also be the case in most developing countries.

Positive net receipts are then related to education. Whether from the perspective of household head's education or from the perspective of spousal education, tendencies are similar. A household with illiterate head or mother is most likely to be net recipient for the reason of illness, while a household with more educated head or mother is more likely to be net recipients for other reasons. This is consistent with research results in the USA, where the probability of both giving and receiving private transfers increases with the level of education (Cox and Raines 1985). The relationship does not appear to hold for Peru (Cox and Jimenez 1992).

Ethnic group is also a significant factor related to being net recipient. Compared with Han people, the minority people are more likely to be net recipients for the reason of sickness and are less likely to be net recipients for other reasons. The household head's occupation is closely related to being net recipient. Farmer is more likely to be a net recipient of gift and money due to illness but less likely to be a net recipient for other reasons, compared to other occupations in rural areas.

Expectedly, households with higher health care costs are significantly more likely to be net recipients due to illness than those with lower health care costs. For other reasons, the transfers are also more likely to flow from the healthy to the unhealthy households. The evidence from other countries such as USA is in support of our results. Schoeni (1993) found both unemployment and poor health increase the probability of receiving private transfers. This pattern was also found in Russia (Cox, Eser, and Jimenez 1997) and in Kyrgyzstan (Cox, Jimenez, and Jordan 1994).

The last factor of interest is household income. The researcher finds that poor households are more likely to be net recipients due to illness than average and rich households. However, average households are more likely to be net recipients for other reasons. Similar results can be found in other researches. The recipients' income is negatively related to the receipt of monetary transfers (Schoeni 1993). In Peru (Cox and Jimenez 1992), the Philippines (Cox and Jimenez 1995), and Russia (Cox, Eser, and Jimenez 1997), lower-income households were more likely to receive transfers. Net private transfer donors usually have greater income than net transfer recipients. In the Philippines, net transfer donors have three times the pre-transfer income of net transfer recipients (Cox and Jimenez 1995). In Russia and Kyrgyzstan, net donors have twice the income of net recipients (Cox, Eser, and Jimenez 1997).

Table 6-19 Description of net gift and cash recipients

		Net recipient due to illness		Net recipient for other reasons	
		Count	%	Count	%
Head sex	Female	9.0	75.0	10.0	83.3
	Male	179.0	62.2	211.0	73.3
Head age	Young	**88.0	59.1	111.0	74.5
	Elder	99.0	66.0	110.0	73.3
Head ethnicity	Han nationality	*149.0	62.1	189.0	78.8
	Yi nationality	39.0	65.0	32.0	53.3
Head education	Illiterate	42	73.7	38	66.7
	Primary education	48	57.1	67	79.8
	Junior high school	45	57.7	56	71.8
	Senior high school	52	65.0	59	73.8
Spousal education	Illiterate	73	71.6	*73	71.6
	Primary education	32	48.5	51	77.3
	Junior high school	33.0	57.9	40.0	70.2
	Senior high school	43.0	64.2	52.0	77.6
Head occupation	Farmer	142.0	64.3	158.0	71.5
	Non-farmer	46.0	58.2	63.0	79.7
Total medical costs	Low	***57.0	46.7	90.0	73.8
	Average	49.0	66.2	52.0	70.3
	High	82.0	78.8	79.0	76.0
Self-rating of comparative wealth	Poor	43.0	68.3	43.0	68.3
	Average	95.0	62.9	117.0	77.5
	Rich	49.0	56.6	60.0	69.5

Source: computed from ZEF survey 2002

Note: *significance at the level of 0.1, **significance at the level of 0.05 and *** significance at the level of 0.01

Effects of the receipts on income redistribution within the extended family

An attempt is made to explore the impact of gift and cash receipts on income redistribution within the extended family by comparing the pre-receipt income with post-receipt income of the sampled households. Annual gift and cash receipt[27] is subtracted from surveyed annual household income, which is taken

[27] Annual gift and cash receipt of each household is obtained by dividing the aggregate

as post-receipt income. Then households by quintile are ranked according to pre-receipt income. However, the public transfers such as social relief for the poorest from the local governments, and the cut-down and exemption of tuition for their kids are not counted in. As shown in Table 6-20, prior to the receipts, the average income of the lowest and the highest quintile is 2656 and 30121 Yuan and accounts for 4.36 and 48 percent, respectively; after the receipts, the share of sum income of the poorest quintile increases by only 0.37 percent while that of the richest decreases by 1.85 percent. The percentage range between the lowest and highest average income shrinks by 2.2 percent. It is interesting to note that the second quintile benefit from the private transfers more than the poorest, indicating that the people in urgent need of assistance cannot acquire private financial aid probably due to their weak financial ability to pay back. An important policy implication is that public programs targeted at the poorest quintile are needed and the effect of crowding out is trivial.

Table 6-20 Effect of receipts on income redistribution within the EF

Income	Pre-transfers		Post-transfers	
quintile	Mean	% of sum	Mean	% of sum
1st	2655.53	4.36	3221.67	4.73
2nd	5592.69	8.59	6683.50	9.81
3rd	9456.37	14.78	10791.94	15.85
4th	15078.87	24.37	16052.36	23.57
5th	30121.58	47.89	31351.63	46.04
Total	12580.17	100.00	13620.22	100.00

Source: computed from ZEF survey 2002

6.2.3 Estimating extended family's impact on health insurance demand

In chapter 5, econometric analyses on the determinants of willingness to join (WTJ) and to pay (WTP) for hypothetical health insurance schemes without taking account of the role of informal insurance arrangements were conducted. However, the neglect of the role of informal insurance arrangements may overestimate demand for formal health insurance. The analysis in previous section exhibited the presence of frequent and regular reciprocal transfers due to illness as well as for other reasons. In consideration the similar function in risk coping between formal and informal insurances, it is anticipated that the presence of informal insurance arrangements has a negative impact on the demand for formal health insurance. Nevertheless, the impact of reciprocal transfers within the extended family on demand for health insurance is two-sided. Apart from smoothing consumption fluctuation, household within the extended family usually adopt active measures to help each other with income generation. It should be pointed out that the transfers recorded may certainly include help for the purpose of income generation. For example, money transfers for payment of seeds, fertilizer or funds of sideline business. Besides

receipts by 15 years.

the capital and labor transfers for production activities, there is also technology and information sharing, which may create more opportunities to earn money for the households within the extended family. Strictly speaking, the extended family, therefore, has both income effect and insurance effect. The income generation may raise the possibility of participating in health insurance schemes (because this study and other studies indicate a positive relation between income and WTJ). Insurance effect of the extended family may substitute to a certain extent for the formal health insurance scheme. The signs of the variables on the function of the extended family depend on the net effect of the two.

Measuring the function of the extended family (risk sharing) is a bit troublesome. Without any existing experience, we have to adopt some tentative measurements and then see which one is preferable. Four ways are utilized to measure the function of the extended family in this study. First, it is measured through the magnitude of gift or cash receipts due to illness (LG_MRECI). We hypothesize that the more money or gift receipts due to illness, the stronger the role the extended family plays in risk sharing other than income earning, so the expected sign is negative. Second, it is measured through the magnitude of gift and cash receipts for other reasons (LG_MRECO). Many of these risks (e.g., housing construction, wedding) are not insurable and have little bearing on health insurance, so this measurement is expected to be less related to the WTJ than the first measurement. Third, the frequency of gift and cash receipt (TIMREC_I) and giving (TIMTRF_I) due to illness is used as measurements. All the previous three are objective measurement of the role of the extended family. Finally, a subjective or perceived measurement is used. For example, relative size of the extended family, it is expected the households belonging to a large extended family has less incentive to participate in RCMS. In consideration of the role of the extended family, we then re-estimate the binary logit model on willingness to join the hypothetical RCMS. Estimation results of four specifications are shown in Table 6-21.

Table 6-21 shows that the first measurement, consistent with our expectation, bears a negative sign, suggesting that the magnitude of gift and cash receipt decrease the WTJ. However, it is not statistically significant. The second measurement bears a positive sign, suggesting that the receipts for other reasons have a strong income effect, relative to insurance effect, and hence boost the participation in the hypothetical RCMS. However, the positive effect is not statistically significant. When the function of extended family is measured by frequency of receipts and donations due to illness respectively, the estimation results are a bit complex. On the one hand, the more frequent one household donates, the less likely they are to be willing to join. On the other hand, the more frequent one household receives, the more likely they are to be willing to join. In this sense, the extended family is more like a risk sharing arrangement, even though the effect is not statistically significant. Where the relative size as a subjective measurement is concerned, contrary to the expectation, the households belonging to a large extended family are more likely to join the hypothetical RCMS.

To sum up, the extended family risk sharing mechanism does not affect the willingness to join the hypothetical RCMS. In other words, the informal insurance arrangement does not resist against formal insurance. It is worthy to note that this should not become the evidence of absence of risk-sharing function of the extended family, since, besides the function of risk sharing, the extended family has an important role in generating income. It is likely that the insurance effect has been offset by the income effect.

Table 6-21 Estimates of the extended family's impact on WTJ: BL model

Dep. variable: WTJ	Coefficient	Std. Error	Coefficient	Std. Error	Coefficient	Std. Error	Coefficient	Std. Error
Indep. variables	(1)		(2)		(3)		(4)	
ETHNICT	2.341***	0.713	2.433***	0.724	3.888***	1.369	2.206***	0.720
HEADAGE	-0.023	0.016	-0.023	0.016	-0.021	0.025	-0.022	0.016
HEADEDU	-0.148***	0.059	-0.149**	0.059	-0.253***	0.083	-0.158***	0.059
SPOUSEDU	0.173	0.056	0.183	0.056	0.268***	0.081	0.183***	0.056
HHSIZE	0.044	0.163	0.028	0.161	0.085	0.224	0.012	0.163
JOBBER	-0.250***	0.255	-0.247***	0.255	-0.998**	0.458	-0.888***	0.321
DEP_IND	-0.892	0.322	-0.928	0.322	0.092	0.376	-0.222	0.256
DISTANS	-0.048	0.041	-0.050	0.041	0.020	0.072	-0.044	0.041
LANDAREA	0.200***	0.058	0.203***	0.057	0.209**	0.080	0.220***	0.059
LG_INCCS	1.214**	0.538	1.188**	0.541	1.758**	0.811	1.209**	0.540
LG_HLCST	-0.014	0.111	-0.039	0.107	-0.062	0.146	-0.026	0.107
CBHISEXP	-0.257	0.350	-0.325	0.353	-0.257	0.459	-0.345	0.353
LG_MRECI	-0.054	0.123						
LG_MRECO			0.059	0.117				
TIMREC_I					0.007	0.014		
TIMTRF_I					-0.020	0.032		
RELASIZE							0.052	0.041
Constant	-2.827	2.178	-2.864	2.182	-5.413	3.302	-2.526	2.621
Number of observations	271		270		166		271	
Log-L	-135.12		-134.61		-73.59		-134.40	
Res. Log-L	-165.29		-164.93		-97.97		-165.29	
Chi-squared	60.34		60.64		48.76		61.78	
Significance	.000		.000		0.000		.0000	

Source: computed from ZEF survey 2002

6.3 Comparison of the extended family with RCMS and PHI

As has been pointed out before, private health insurance (PHI) policies and rural cooperative medical schemes (RCMS) may just crowd out the existent informal insurance mechanisms in rural areas, and hence no welfare gain is obtained from the introduction of the formal health insurance. Thus, a careful comparison between formal and informal mechanisms is needed to judge the potential welfare gain or loss. Methodologically, county filiales of China life insurance company, RCMS at the town level and the extended families are compared in terms of institution, technique and management.

Without the loss of representativeness, the researcher surveyed 5 county filiales of China life insurer that supplies major health insurance products in rural areas, 7 rural cooperative medical schemes at the town level[28]. The data on 300 extended families are available on the basis of HISRS. All the quantitative indicators describing attributes of the three types of organizations are mean of its respective samples. The comparison is summarized in Table 6-22.

Despite their common function in risk coping, the extended family as a risk-sharing arrangement differs substantially from commercial life insurers and rural cooperative medical schemes. In terms of institution, the extended family is an informal organization without legislative basis, while the life insurer is a quite formal private company, established on the basis of a series of laws: corporate law, insurance law and other related laws and regulations. The community based health insurance scheme is less formal than the life insurer. It is initiated based on the regulations of health authorities. The life insurer can be regarded as a risk-neutral organization and supplies many kinds of life-related insurance products, health insurance being only part of their businesses. RCMS mainly deals with health risk and hence offers only single product, while the extended family can be thought of providing a comprehensive mutual insurance product because this group deals not merely with health risk, but also with other forms of risks.

The three organizations also differ in objectives. As a private corporate, the life insurer is certainly in pursuit of profit, and RCMS is a non-profit organization aimed at welfare improvement of the locals, while the extended family intends to resist against multifold risks through reciprocal transfers and hence smoothes consumption. Both PHI organization and RCMS stipulate the relation between the insurer and the insured by contract; any party that fails to fulfill its obligations can be taken to court. On the contrary, the relations between the household members of the extended family are quite lax at first glance, but in fact they need not have more problems with the enforcement, because the sanction of exclusion from the group (autarky) may be rather undesirable in the setting of rural China, where farm activities are under-mechanized and insurance market is missing. Also, they have different incentives. Life insurers try to maximize their profits and RCMS to serve rural people, whereas the extended family is aimed to meet social norms and values and show their altruism and charity (actually the behavior does not contradict self-interest). Concerning the target population, the life insurers sell insurance products to all, and RCMS provides health care benefits only for the local rural residents, while the extended families confine themselves to the member

[28] It was originally planned to survey two rural cooperative medical schemes in each county, which is not necessarily in the sampled towns. However, Yanyua county has no RCMS at all and Fushun county has only one pilot RCMS. In total seven rural cooperative medical schemes were investigated.

households within the extended families. Due to the regular and frequent interaction, member households within the extended family have sufficient information exchange so as to be able to deal easily with the problem of moral hazard. In contrast, the life insurers and RCMS are subject to the problems of moral hazard and adverse selection, that is, the transaction costs of the formal organizations are higher than that of the informal one.

In terms of technique and management, the formal and informal organizations are also distinct. The managerial expense of the life insurer is large but not specific for health insurance because the health insurance is not independent in reckoning. Many managers of insurance companies believed that the managerial expenses of health insurance were much larger than those of other insurance policies; the managerial expense of the extended family is unclear but small. However, the managerial expense of RCMS is fixed at the level of 3 percent. The executives of two formal organizations are managers, directors of township health centers, respectively. The extended family does not have any explicit managers. Implicitly, the richest member households (or the most respected one in the extended family) act as the managers. In terms of size of risk pool, the life insurer has substantial advantages. The life insurer has an average of 62,189 health insurants, and RCMS has an average of 10,420 members (or 2,977 households). In contrast, the extended family is fairly small, with about 40 persons (or 10 households including the respondent's household) sharing risks. The premiums of insurance policies with private insurers vary from 30 to 1500 Yuan per year; RCMS requires the members to contribute 6 to 10 Yuan per capita per year. The extended family, nevertheless, does not have an explicit concept of premium, the member households share risk with part of their assets if incomplete insurance is assumed. The forms of benefits of the three are also different. The life insurer and RCMS take the form of cash or the insured get health care directly with the third party paying for the care. The benefits in extended family take the forms of cash, gift or labor exchange. The various forms of benefits make member households more flexible to carry out the obligations. Although the risk pool of the extended family is relatively small, the magnitude of the compensation is not the case. The average compensation of life insurer for those who have fallen ill is 607 Yuan and that of RCMS 116 Yuan. In contrast, the compensation for the sick from their relatives and friends amounts to 206 Yuan, exceeding that of RCMS. This amount is quite closer to the average amount of hospitalization costs (240 Yuan) in Sichuan province. Moreover, the extended family has other merits such as freedom in selecting doctors and the quickness of compensation. However, the compensation has a large uncertainty.

Table 6-22 Comparison of the extended family with PHI organization and RCMS

	China Life Insurer at county level (n=5)	RCMS at township level (n=10)	Extended family (n=300)
Institution and organization attributes			
Form of	Private	Public-private collective	Informal organization

organization		organization	
Legislative basis	Corporate law and insurance law, etc.	The regulation of health sector or related sector	No
Insurance products	Many	Single	"A informal comprehensive mutual insurance"
Objective	Profit	Welfare, non-profit	Resistance against multiple risks
Relation between insurer and insured	Formal contract relation, strong sanction for defection	Informal contract relation, less strong sanction than the former	No contract, but the blood-relation & inherent informal family agreement
Incentives	Maximization of profit	Mutual help, solidarity and serving rural people	Meeting social norms and value, altruism, charity; self-interest
Sanctions	The guarantee corresponds with the premium level	Social pressure, even exclusion from the organization	Inherent family contract, social pressure
Target population	All the locals	Local rural residents	Relatives plus good friends
Information	Asymmetric	Less symmetric	Symmetric
Moral hazard & adverse selection	Severe	Slight	Very slight
Technical and management attributes			
Managerial expense	High but unspecific	3% of premium income	Very low or negligible
Executives	Managers of county branch companies	Directors of town health centers	Implicitly, the rich or the most respected relatives
Pool size	62188.80 persons	2977 hh /10420 persons	10 households/40 persons*
Premium (Yuan)	30-1500 per capita per year	6-10 per capita per year	No explicit concept of premium, risk sharing with part of household assets
Premium income	3717590.60 Y	62217.50 Y	-
Benefit forms	Cash	Cash	Gift, cash, labor, etc.
Persons claimed in 2001	2049	597	15
Total compensation	1243835.60 Y	69379.90 Y	3102.3 Y
Compensation rate**	106.68%	85.00%	-
Average compensation	**607.50** Y	**116.21** Y	**206.82** Y
Hospitals designated	County hospitals and above	Town hospitals and above	No restriction
Convenience of seeing doctor	Far and inconvenient	Near and relatively convenient	Convenient or it depends
Requirements for repayment	Formal invoices in county hospitals	Invoices and receipts in public facilities	No specific requirements

Source: ZEF survey 2002

Note: *Here the extended family includes the respondent's household itself. If good friends are not counted in extended family, the average size is 6.8 households. **Compensation rate is calculated by dividing the compensation by the premium income.

6.4 Summary and recommendations

Although the direct illness costs in rural China, including financial and time costs of illness, are moderate in comparison with other developing countries such as Ethiopia and Burkina Faso, its share in household income (29%) is high sufficiently to adversely affect normal food consumption and may induce poverty under some circumstances. However, whether or not the costs of illness induce poverty depends to a large extent upon the capacity of coping with the costs. The household is the basic unit of risk coping, 72 percent of illness episodes are successfully financed within households. The extended family plays an important role in coping with relatively large costs of illness, 22 percent of illness episodes had to seek help from other households within the extended family. The remaining illness episodes may be catastrophic and demand aids from communities or governments.

The role of the extended family in coping with costs of illness was highlighted in this chapter. First, its basic attributes were measured. The extended family is composed of an average of ten households of relatives (including the respondent's household). The most important donors to the respondent's household were brothers and sisters who accounted for more than half of the relatives and friends. The majority of the donors were at least not poorer than the respondent's households. Compared with the respondent's households that all live in rural areas, 16 percent of their relatives or friends live in urban areas; the relatives who offered financial aid were more likely to be males. Second, the magnitude of private transfer was measured. On average, each relative donated 2.4 times of gift and cash (a worth of 315.6 Yuan) to the respondents' household due to illness; in total, the respondents' households received 16 times of gifts and cash (a worth of 1627.1 Yuan) due to illness. Of the respondents' households, 63 and 74 percent were net recipients due to illness and for other reasons, respectively. Whether or not a household was a net recipient due to illness was related to gender, age, ethnic group, and education of the household head, and also the medical costs and income level of the household.

Though the extended family plays an important role in helping the unfortunate households pass through hardships, the effect of private transfers on income re-distribution is mild. We further find that most private transfers were captured by the second poorest quintile rather than the poorest. This result suggests that the public program targeted at the poorest quintile has scarcely an effect of crowding out private transfers. The frequent and small-sized private transfers clearly demonstrate the existence of risk sharing within the extended family. Also, the function of extended family in income generation was displayed. Belonging to larger or financially stronger extended families, households were more likely to purchase a private health insurance policy or be willing to join and pay for a hypothetical RCMS. However, the study did not find any evidence in support of the extended family crowding out private health insurance and RCMS. It is worthwhile to note that the insignificant net effect of the extended family on the demand for formal insurance should not become evidence in support of the absence of risk sharing mechanism within the extended family. Finally, by making organizational, institutional and technical comparisons between the extended family, private health insurance organization and RCMS, we find out that the extended family has much lower transaction costs than the formal insurance arrangements. Moreover, a fatal deficiency of the existing RCMS in Sichuan is that the coverage is even lower than the sum of private transfers from the extended family.

A handful of policy suggestions can be extrapolated from the results of this study. To enrich the rural households' strategies for coping with risk and strengthen the self-enforcement of the reciprocal private transfer within the extended family, mutual help and altruism should be widely propagandized and encouraged by the public voice. In the process of designing RCMS, the executive agencies should set a suitable premium to be able to cover both outpatient and inpatient services in light of the magnitude of financial costs of illness. To attract the rural households into joining RCMS, the indemnity of RCMS should be higher than the sum of private transfers due to illness within the extended family. Meanwhile, some public programs targeted at the poorest are necessary and feasible. Besides the household income, the scale of extended family, especially the number of brothers and sisters, facilitates targeting the intended population group.

7. General Conclusion and Recommendation

7.1 Main findings

This dissertation provides an empirical study on health insurance demand and health risk management in rural China on the basis of two surveys: the Health Insurance Survey in Rural Sichuan (HISRS) and the National Survey on Rural Health System (NSRHS). Not only have demand for private health insurance (PHI) and rural cooperative medical scheme (RCMS) and their impact on care-seeking behavior been described and quantitatively analyzed, but also the role of the extended family as an informal risk-sharing arrangement and a major risk coping mechanism in China's social, economic and cultural setting has been measured and analyzed. The findings associated with the three forms of health care financing—PHI, RCMS and the extended family in rural China are summarized separately as follows.

Private health insurance (PHI)

The study tries to explore whether the private health insurance can benefit rural people and how it can be done by investigating who are privately insured and how their care-seeking behaviors are affected by the ownership of PHI policies. Results indicate that about 12 percent of people in the sample were policyholders of the PHI; the largest group of beneficiaries of private health insurance is in school, from kindergarten kids to college students who usually participate on a group basis. Using a binary probit model, we find that despite many determinants, purchasing PHI is individual specific and varies largely across counties in Sichuan Province. Age, education of individual and relation to the household head are all significantly related to the purchase of PHI. Adverse selection does exist as anticipated but it is mild. Moreover, demand for PHI is related to the effort of the supply side. Where there is fierce competition between insurance suppliers, there is a high purchase rate.

On the basis of the Sichuan dataset, both descriptive and econometric analyses demonstrate that PHI does not have a significant impact on illness reports and health care seeking decisions after controlling for income and demographic attributes. PHI is not conducive to active illness reports and to quick care-seeking decisions as expected. Additionally, the premium is relatively high. However, it affects the choice of providers by encouraging policyholders to seek care in public health facilities (village health stations, town health centers and county hospitals) rather than private ones, especially to seek high-quality (efficacious) health care that is provided by county hospitals. Thus, it still benefits the rural households by insuring the children (usually one child of one couple) against costs of serious diseases and injuries with an insurance product called 'pupil health and accident insurance'. This suggests that though PHI has a lot of problems such as the discrimination against the high-risk population and high transaction costs, it remains a useful health care financing instrument in rural areas.

In the process of investigating the impact of PHI on health care seeking behavior, a by-product is yielded by computing the income and price elasticity of four providers in a multinomial logit model. Results show that health care in general is a normal good and a necessity; in particular, health care provided by town health centers and village health stations is an inferior good and health care provided by county hospitals and private facilities is a normal good. The health services provided by county hospitals and private facilities are much more price responsive than those provided by the two others. Compared with the non-poor, the poor are more income responsive to the most and least desirable health services (from county hospitals and village health stations, respectively). This suggests that the proper difference in quality has its rationality and the inferior goods can be beneficial to poor people while naturally excluding the rich. The poor are also more price elastic to health care provided by private facilities and county hospitals.

Rural cooperative medical schemes (RCMS)

Much attention was paid to the rural cooperative medical system to assess its huge potential in resource mobilization and financial protection and to come up with approaches to the sustainable development of RCMS. Not only were the observed demand for the RCMS and its performance explored and assessed, but also the potential demand for a hypothetical RCMS was investigated and computed. In the study on real demand, we find that about 16 percent of the sampled households were members of RCMS. However, this level should not be taken as the general level of enrolment rate of RCMS in Sichuan since RCMS is not uniformly distributed. A descriptive analysis indicates that the deep involvement of the local government and its fiscal subsidies to the scheme are crucial to the establishment and persistent operation of RCMS at the level of village or town. Furthermore, collecting premium on a group basis in combination with payment of agricultural taxes and fees makes transaction costs low despite the prohibition by the Ministry of Agriculture. Literature reviews and our own case study indicate that RCMS in China has different capacities of resource mobilization and financial protection in different regions. The eastern developed region not only has a high enrolment rate but also mobilizes a large amount of resources from governments and local communities, whereby RCMS can protect its members well against the costs of illness. In contrast, the middle and western regions have a low enrolment rate, and a weak capacity of resource mobilization and financial protection. An attempt is made to link RCMS to health improvement. The comparative analyses on the basis of a national data set collected by Chinese Academy of Social Science and statistics data from National Statistic Bureau show that RCMS as a whole played a big role in mobilizing resources and hence enhancing the national health status before the economic reforms. After the economic reforms RCMS appears to have no significant impact on the health status improvement at county level due to the low coverage.

In the study on potential demand for a hypothetically improved RCMS, the contingent valuation (CV) approach was used to create a hypothetical market for an improved RCMS with complete benefits and strict fund management. Though the CV approach is not without problems, it remains a useful starting point for the analysis of the extension of RCMS. Findings show that the willingness to join the hypothetical RCMS is not particularly high (69%) in consideration of the high enrolment rate (over 90%) of RCMS in the 1960s and 1970s. For those respondents who are unwilling to join, the distrust in the management of insurance fund is the most frequently cited reason (37%) and the non-affordability of the poor ranks second (28%). Among the three hypothetical benefits packages of RCMS are inpatient services coverage with no copayment, complete coverage with half co-payment and complete coverage with no co-payments. The second benefits package is most desirable because the mean willingness to pay (WTP) for it (32.50 Yuan) exceeds half the WTP for complete coverage without co-payment (58.08 Yuan) and is much higher than the premium of the present RCMS (6 Yuan). The hypothetical RCMS with half co-payment distinguishes itself from the existing RCMS by the hypothetically strict management of the insurance fund and enlarged benefits (both out- and inpatient services). This clearly demonstrates that to enhance the willingness to join and pay, a RCMS with a stronger capacity of financial protection and a stricter insurance fund management must be designed. RCMS is a pro-poor instrument because it has a favorable impact on health care seeking behavior. On the basis of the Sichuan dataset, a descriptive analysis demonstrates that RCMS significantly encouraged the members to report their sickness (even minor sickness) actively and seek care in village health station and town health centers. The regression analyses of the respondents' WTJ and WTP show that being minor ethnic group and having higher income are two strong boosting factors whilst having a migrant worker is a significantly negative determinant.

Extended family and risk coping

Besides the two formal insurance arrangements, this study also tries to investigate the strategies of households for coping with health risk in general and the role of the extended family in particular. The extensive strategies for coping with economic costs of illness were investigated before we turned to the extended family. Results reveal that, though the economic costs of illness, inclusive of monetary and time costs of illness, are moderate, compared with other developing countries such as Ethiopia and Burkina Faso, the share of economic costs of illness in household income (29%) is sufficiently high to adversely affect basic food consumption, and may induce poverty under some circumstances. Whether or not the costs of illness induce poverty depends to a large extent upon the household's capacity of coping with the costs. Findings show that, consistent with previous studies, household is the basic unit of risk coping. Seventy-two percent of illness episodes were successfully financed within households through a variety of strategies such as cash, savings, sale of agricultural products, livestock or other assets. The extended family played an important role in coping with costs of illness, with 22 percent of the illness episodes seeking financial aid from the extended family.

The study highlights the role of the extended family in coping with costs of illness. First, its basic attributes were measured. The extended family has an average of ten households (including the respondent's household). The most important donors to the respondent's household are brothers and sisters who account for more than half of the relatives and friends. The majority of the donors are at least not poorer than the respondent's households. Compared with the respondent's households that all live in rural areas, 16 percent of their relatives live in urban areas; the relatives who offered financial aid were more likely to be males. Second, the magnitude of private transfer was measured. On average, each relative donated 2.4 times (a worth of 315.6 Yuan) to the respondents' household due to illness; in total, the respondents' households received 16 times (a worth of 1627.1 Yuan) due to illness. Of the respondents' households, 63 and 74 percent were net recipients due to illness and for other reasons, respectively. Whether or not a household was a net recipient due to illness was related to sex, age, ethnic group, and education of the household head, and also the medical costs and income level of the household.

Though the extended family plays an important role in helping the unfortunate households pass through hardship, the effect of private transfers on income re-distribution is mild. We further find that most private transfers were captured by the second poorest quintile rather than the poorest. This result suggests that the public program targeted at the poorest quintile has scarcely an effect of crowding out private transfers. The frequent and small-sized private transfers clearly demonstrate the existence of risk sharing within the extended family. Also, the function of extended family in income generation is displayed. Belonging to a larger or financially stronger extended family, a household was more likely to purchase a private health insurance policy or be willing to join and pay for a hypothetical RCMS. However, the study has not found any evidence that the extended family crowded out private health insurance and RCMS. It is worthwhile to point out that the insignificant net effect of the extended family on demand for formal insurance should not become evidence in support of the absence of risk sharing mechanism within the extended family. Finally, by making organizational, institutional and technical comparisons between the extended family, private health insurance organization and RCMS, we find out that the extended family has much lower transaction costs than the formal insurance arrangements. Moreover, a fatal deficiency of the existing RCMS in Sichuan is that the coverage is even lower than the sum of private transfers from the extended family.

7.2 Policy recommendation

A handful of policy recommendations can be extrapolated from the results of this study. Though both RCMS and PHI are promising health care financing instruments in rural areas, RCMS is a more pro-poor tool and deserves to be translated into instant action by the communities and governments.

To raise the purchase rate of PHI, the insurers should give priority to the younger, better-educated and risk-averse individuals in richer households that have a large portion of cash income. Meanwhile, the governments need to encourage commercial insurers to be more deeply involved in the rural market. Three ways should be given the priority. One is to cut tax rate of some products targeted at the vulnerable groups (e.g., the kids and pupils) that governments want to subsidize; another is to invest more in the preventive and primary care for the rural population to reduce their health risk; the other is to effectively regulate and supervise the town health centers so that they are eligible as the designated hospitals of the insurance companies and hence decrease costs of care as well as transaction costs. In addition, governments need to foster a competitive insurance market.

To expand the demand for RCMS, the governments and communities need to enlarge RCMS's capacity of mobilizing resources, and of protecting its members against costs of illness. To fulfill this task, the executive agencies should set a suitable premium to be able to cover both outpatient and inpatient services. In more detail, RCMS should be designed to have a complete coverage with half co-payment, a private contribution of around 30 Yuan per capita per annum, a strict management of insurance funds and enhanced subsidies from local and central governments, especially for rural people in underdeveloped middle and western provinces. To be more attractive for the rural households, RCMS should pay a higher compensation than the sum of private transfer due to illness within the extended family. Different levels of government should play different roles. The central government should act as cross-province subsidizer and legislator; the provincial and county governments should play an active role in subsidizing, guiding and regulating RCMS; and the town governments should be major executive agencies and administrators while the villages serve to propagandize RCMS towards rural people face-to-face and to collect premiums at low costs. The minority communities with middle income can be prioritized if they have possessed the basic capacity of implementing and managing RCMS. However, the issue of high proportion of migrant worker should be properly addressed, otherwise the hope of reconstructing RCMS in rural areas is dim.

Concerning the extended family, the local officials and village cadres have a role to play in strengthening the self-enforcement of the reciprocal private transfers. The prohibition of constructing house on the land of other households can maintain the proximity of housing, which in turn increase informational symmetry among the relatives and decrease the likelihood of reneging on commitments. The condemnation of failure of commitment by the village cadres can also increase the costs of defection. Meanwhile, some public programs targeted at the poorest are necessary and feasible. Besides the household income, the information on the scale of the extended family, particularly on the number of brothers and sisters, facilitates targeting the intended population group. More efforts to collect this baseline information should be funded.

To provide adequate health care for the rural poor, the government should adopt effective measures to raise the share of cash income in total household income. Village health station and town health centers need to be better equipped and staffed with the assistance of public health investment and RCMS fund. However, a suitable difference in quality is required to ensure that the rural poor have easy access to cheap health care.

In conclusion, both the private health insurance products and rural cooperative medical schemes, in conjunction with the informal insurance arrangements, particularly the extended families, should be combined to provide rural people with an effective health risk protection in consideration of the relatively low productivity and the huge regional differences in China today. However, the focus of governments and communities should be placed on the construction and sustainable development of RCMS. The large-scale government subsidies will undoubtedly raise demand for rural cooperative medical scheme, but identifying the optimal level of governmental subsidies and analyzing the effect of the governmental subsidies on demand for RCMS deserves further attention in the future research.

Reference

Abel-Smith, B. and A. Dua. 1987. The potential of community financing of the health sector in developing countries. In *Health Care Financing*. Asian Development Bank: 42-70.

Abreu, D. 1988. *On Theory of Infinitely Repeated Games with Discounting*. Econometrica. 56: 282-96.

ADB. 2002. *Asian Development Outlook 2002*. Oxford, UK: The Oxford University Press.

Akerlof, G.A. 1970. The Market for "Lemons": Quality Uncertainty and the Market Mechanism. *Quarterly Journal of Economics*. 84(3): 488-500.

Amemiya, T. 1981. Quantitative Response Models: A Survey. *Journal of Economic Literature*. 19(4): 481-536.

Arhin-Tenkorang, D. 2001. Health Insurance for the Informal Sector in Africa: Design Features, Risk Protection, and Resource Mobilization. CMH Working Paper Series No. WG3: 1/2001. WHO: Commission of Macroeconomics and Health (CMH).

Arnott, R. 1990. Moral Hazard and Competitive Insurance Markets. In *Foundation of Insurance Economics*. Dionne, G., and S. E. Harrington. 1991. Dordrecht, the Netherlands: Kluwer Academic Publisher.

Arnott, R and J.E. Stiglitz. 1990. The Welfare Economics of Moral Hazard. In *Foundation of Insurance Economics*. Georges, D. and S.E. Harrington. 1991. Dordrecht, the Netherlands: Kluwer Academic Publisher.

Arrow, K.J. 1963. Uncertainty and the Welfare Economics of Medical Care. *American Economics Review*. 53: 941-969.

Arrow, K.J. 1965. Insurance, Risk and Resource Allocation. In *Aspects of the Theory of Risk Bearing*. Arrow K.J. (eds.). Helsinki, Finland: Yrjö Jahnssonin Foundation. Reprinted in *Essays in the Theory of Risk Bearing*. Arrow, K.J. 1971: 134-143. Elsevier Publishing Company Inc.

Arrow, K.J. 1996. The Theory of Risk-Bearing: Small and Great Risk. *Journal of Risk and Uncertainty*. 12: 103-111.

Arrow, K., R. Solow, P. Portney, E. Leamer, R. Radner, and H. Schuman *et al.* 1993. *Report of NOAA Panel on Contingent Valuation*. Federal register 58 (1993): 4601-14.

Asenso-Okyere, W.K, I. Osei-Akoto and A. Anum *et al.* 1997. Willingness to Pay for Health Insurance in a Developing Economy: A Pilot Study of the Informal Sector of Ghana Using Contingent Valuation. *Health Policy*. 42: 223-237.

Asfaw, A. 2002. *Costs of Illness, Demand for Medical Care, and the Prospect of Community Health Insurance Schemes in the Rural Areas of Ethiopia*. Frankfurt am Main, Germany: Peter Lang, Europaeischer Verlag der Wissenschaft.

Atim, C. 1999. Social Movements and Health Insurance: A Critical Evaluation of Voluntary, Non-profit Insurance Schemes with Case Studies from Ghana and Cameroon. *Social Science & Medicine*, 48(7): 881-896.

Behrman, J.R., A. Foster, and M. R. Rosenzweig. 1997. Dynamic Savings Deci- sions in Agricultural Environments with Incomplete Markets. *Journal of Business and Economic Statistics*. 15(2): 282-292.

Berman, P.A. 1995. Health Sector Reform: Making Health Development Sustainable. In *Health Sector Reform in Developing Countries: Making Health Development*

Sustainable. Berman, P.A. (eds.): 14-33. Boston: Harvard University Press.

Binswanger, H. and J. McIntire. 1987. Behavioral and Material Determinants of Production Relations in Land-Abundant Tropical Agriculture. *Economic Development and Cultural Change*. 36(1): 73-99

Bishop, R.C. and T.A. Heberlein. 1979. Measuring Values of Extra-Market Goods: Are Indirect Methods Biased? *American Journal of Agricultural Economics*. 61: 926-930.

Bloom, G. and D. McIntyre 1998. Towards Equity in Health in An Unequal Society. *Social Science and Medicine*. 47(10): 1529-1538.

Bogg, L., H. Dong, K. Wang, W. Cai, and V. Diwan. 1996. The Cost of Coverage: Rural Health Insurance in China. *Health Policy and Planning*. 11(3): 238-252.

Butz, W. and P. Stan. 1992. Household Composition and Inter-household Exchange in Malasia. *Population and Development Review.* 8(3): 92-115.

Cai, F. and Y. Dou. 2001. *China Report on Population issues: Education, Health and Economic Growth*. Beijing, China: Social Science Documentation Publisher.

Cameron, T.A. and M.D. James. 1987. Efficient Estimation Methods for "Closed-ended" Contingent Valuation Surveys. *Review of Economics and Statistics*. LXIX(2): 269-276.

Cameron, T.A. 1988. A New Paradigm for Valuing Non-market Goods Using Referendum Data. *Journal of Environmental Economics and Management*. 15: 355-379.

Carrin, G., D.D. Graeve and L. Deville. 1999. Introduction to Special Issue on the Economics of Health Insurance in Low and Middle-income Countries. *Social Science and Medicine*. 48: 859-864.

Caroll, C.D. 1992. The Buffer-Stock Theory of Saving: Some Macroeconomic Evidence. *Brookings Papers on Economic*. Activity, 2: 61-156.

Carson, R.T. 1989. *Constructed Markets*. Mimeo.

Carson, R.T. 1997. Contingent Valuation: Theoretical Advances and Empirical Tests Since the NOAA Panel. *American Journal of Agricultural Economics*. 79: 1501-1507.

Carson, R. T. 2000. Contingent Valuation: A User's Guide. *Environmental Science and Technology*. 34: 1413-18.

Carson, R.T, L. Wilks and D. Imber (1994). Valuing The Preservation of Australia's Kakadu Conservation Zone. *Oxford Economic Papers*. 46: 727-749.

Carson R.T., G. Theodore and M.J Machina. 1999. *Incentive and Informational Properties of Preference Questions*. Paper presented at the IX Conference of the European Association of Environmental and Resource Economists. Oslo 25-27 June, 1999.

Carson, R.T., N.E. Flores, and N.F. Meade. 2000. Contingent Valuation: Controversies and Evidence. *Forthcoming Environmental and Resources Economics*. Mimeo.

Chakrabarti, S., W. Lord, and P. Rangazas. 1993. Uncertain Altruism and Investment in Children. *American Economic Review*. 83(4): 994-1002.

Chawla, M. and R. Pellis. 2000. The Impact of Financing and Quality Changes on Health Care Demand in Niger. *Health Policy and Planning*. 15(1): 76-84.

Chen, J., Z. Lüand Y. Wang. (Eds.) 2001. *The Development Report of China's Social Security System (1997-2001)*. Beijing, China: The Social Science Documentation Publisher.

Chen, M.A.. 1991. *Coping with seasonality and Drought*. Sage, New Deli.

Chen, N., A. Ma and Y. Guan. 2000. *Study and Experience of a Risk-based Cooperative Medical System in China: Experience in Weifang of Shandong Province*. Paper for

International Conference on Health System Financing in Low-Income African and Asian Countries, December.
Chen, X., T. Hu and Z. Lin. 1993. The Rise and Decline of the Cooperative Medical System in Rural China. *International Journal of Health Services.* 23(4): 731-742.
Chiu, W.H.. 1997. Health Insurance and the Welfare of Health Care Consumers. *Journal of Public Economics.* 6: 125-133
Coate, S and M. Ravallion. 1993. Reciprocity without Commitment, Characterization and Performance of Informal Insurance Arrangements. *Journal of Development Economics.* 40: 1-24.
Coleman, J.S. 1994. Social Capital in the Creation of Human Capital. In *Social Capital: A Multifaceted Perspective.* Partha Dasgupta and Ismail Serageldin (eds.) 1999. Washington, DC: The World Bank.
Costa, J. and J. Garcia. 1999. Demand for Private Health Insurance: Is There A Quality Gap. http://ideas.uqam.ca. Accessed in June 2001.
Cox, D. and E. Jimenez. 1990. Achieving Social Objectives through Private Transfer: A Review. *The World Bank Research Observer.* 5(2): 205-18.
Cox, D. and E. Jimenez. 1992. Social Security and Private Transfer in Developing Countries: The Case of Peru. *World Bank Economic Review.* 6(1): 155-169.
Cox, D., E. Jimenez and J. Jordan. 1994. Family Safety Nets and Economic Transition: A Study of Private Transfers in Kyrgyzstan. Mimeograph.
Cox, D. and E. Jimenez. 1995. Private Transfer and the Effectiveness of the Public Income Redistribution in the Philippines. In *Public Spending and the Poor: Theory and Evidence.* Van der Walle, D. and K. Nead. (eds.): 321-345. Published for the World Bank. Baltimore and London: The John Hopkins University Press.
Cox, D. and F. Raines. 1985. Inter-Family Transfer and Income Redistribution. In *Horizontal Equity, Uncertainty, and Economic Well-being.* David, M. and S. Timothy. (eds.). Chicago, US: University of Chicago Press.
Cox, D., Z. Eser and E. Jimenez. 1997. Family Safety Nets during Economic Transition. In *Poverty in Russia: Public Policy and Private Responses.* Klugman, J. (ed.): 211-248. EDI Development Study.
Daniels, L. 1994. *Changes in the Small-Scale Enterprise Sector from 1991 to 1993: Results from A Second Nationwide Survey in Zimbabwe.* Gemini Technical Report No. 71. Gemini, Bethesda, Maryland.
Davis, S. 1993. Are Coping Strategies A Cop Out? *IDS Bulletin.* 24: 60.
Deaton, A. 1997. The Analysis of Household Surveys: A Microeconomic Approach. Baltimore, Md.: Johns Hopkins University Press.
Dercon, S. 2002. Income Risk, Coping Strategies, and Safety Nets. *The World Bank Research Observer*, 17(2): 141-166.
Dercon, S. and P. Krishnan. 1996. Income Portfolios in Rural Ethiopia and Tanzania: Choices and Constraints. *Journal of Development Studies.* 32(6): 850–875.
Dercon, S. and P. Krishnan. 2000. In Sickness and in Health: Risk Sharing within Households in Rural Ethiopia. *Journal of Political Economy.* 108(4): 688-727.
Dickerson, O. D. 1959. *Health Insurance.* Ed. Richard D. Irwin, Inc.
Diener, A., B. O'Brien, A. Gafni. 1998. Health Care Contingent Valuation Studies: A Review

and Classification of the Literature. *Health Economics*. 7: 313-326.

Dionne, G. and S. E. Harrington. 1991. *Foundation of Insurance Economics*. Dordrecht, the Netherlands: Kluwer Academic Publisher.

Dong, H. 2000. *Health Financing System & Drug Use in Rural China*. Stockolm, Sweden: IHCAR, Karolinska Institut, SE-171 76.

Dror, D., and C. Jacquier. 1999. Micro-insurance: Extending Health Insurance to the Excluded. *International Social Security Review*. 52(1): 71-97.

Fafchamps, M. 1992. Solidarity Networks in Pre-industrial Society: Rational Peasants with A Moral Economy. *Economic Development and Cultural Change*. 41(10): 147-73.

Fafchamps, M. 1999. *Rural Poverty, Risk and Development*. FAO Economic and Social Development Paper No. 144. Rome, Italy: Food and Agricultural Organization.

Fafchamps, M. and Gavian S. 1997. The Determinants of Livestock Prices in Niger. *Journal of African Economies*. 6(2): 255-295.

Fafchamps, M. and S. Lund. 2002. Risk Sharing Networks in Rural Philippines. *Journal of Development Economics* (forthcoming).

Fafchamps, M. and A. Quisumbing. 1999. Human Capital, Productivity, and Labor Allocation in Rural Pakistan" *Journal of Human Resources*. 34 (2): 369-406.

Fafchamps, M. and A. Quisumbing. 2003. *Social Roles, Human Capital, and the Intra-household Division of Labor: Evidence from Pakistan*. Oxford Economics Papers, 2003 (forthcoming).

Farmer, P. 1999. *Infections and Inequality: The Modern Plagues*. Berkeley, US: University of California Press.

Feldstein, M. The Welfare Loss of Excess Health Insurance. *Journal of Political Economy*. 81: 251-280.

Folland, S., A. Goodman, and M. Stano. 2001. *The Economics of Health and Health Care*. New Jersey, US: Prentice-Hall, Inc., Upper Saddle River.

Andrew, D., M. Foster and R. Rosenzweig. 1996. Comparative Advantage, Information and the Allocation of workers of Tasks: Evidence from An Agricultural Labor Market. *Review of Economic studies*. 63(3): 347-374.

Frank, R. 1999. *Microeconomics and Behavior*. New York, US: Mcgraw Hill Inc.

Gale, W. and J. Scholz. 1993. Intergenerational Transfer and Wealth Accumulation. Discussion Paper No. 1019-93. Institute for Research on Poverty, University of Wisconsin-Madison.

Gao, G., X. Han, and S. Xu *et al*. 2002. Survey and Analysis of Funding and Compensation of the Rural Health Care Scheme in Shanghai. *China's Rural Health Administration*. 22 (2): 45-48. In Chinese.

Gao, J., J. Qian, S. Tang, B. Eriksson and E. Blas. 2002. Health Equity in Transition from Planned to Market Economy in China. *Health Policy and Planning*. 17(*Suppl*. 1): 20-29.

Gertler, P. and J. Gruber. 1997. Insuring Consumption Against Illness. NBER Working Paper No. 6035. Health Care and Public Economics.

Gertler, P. and O. Solon. 2000. *Who Benefits From Social Health Insurance in Developing Countries*? Paper for International Conference on Health System Financing in Low-Income African and Asian Countries, December.

Gertler, P. and J. van der Gaag. 1990. *The Willingness to Pay for Medical Care: Evidence from Two Developing Countries*. Baltimore, Md.: The Johns Hopkins University Press.

Green, D., K. Jacowitz, K. Daniel and M. Daniel. 1998. Referendum Contingent Valuation, Anchoring, and Willingness to Pay for Public Goods. *Resource and Energy Economics*. 20(2): 85-116.

Greene, W. 2000. *Econometric Analysis*. Fifth Ed., New York, US: Macmillan Publishing Company.

Grossman, M. 1972. On the Concept of Health Capital and the Demand for Health. *Journal of Political Economy*. 80: 223-255

Gujarati, D. 1994. *Basic Econometrics*. 3rd Ed. New York, US: Mcgraw-Hill Inc.

Gu, X. and S. Tang. 1995. Reform of Chinese Health Care Financing System. *Health policy*. 32: 181-191.

Gu, X. and S. Xue. 2000. *Study on Health Financing in Rural China*. Paper for International Conference on Health System Finance in Low-Income African and Asia Countries.

Hammer, J. and P. Berman. 1995. Ends and Means in Public Health Policy in Developing Countries. *Health Policy*. 32: 29-45.

Hamoudi, A. and J. Sachs. 1999. Economic Consequence of Health Status: A Review of the Evidence. CID working Paper. No. 30. Cambridge, Massachusetts: Center for International Development at Harvard University.

Hanemann, W. 1989. Welfare Evaluation in Contingent Valuation Experiments with Discrete Response Data: Reply. *American Journal of Agricultural Economics*. 71: 1057-61.

Hardaker, J., R. Huirneand and J. R. Anderson. 1997. *Coping with Risk in Agriculture*. Wallingford, UK: Commonwealth Agricultural Bureaux (CAB) International.

Hayami, Y. 1997. *Development Economics: from Poverty to Wealth of Nations*. Oxford, UK: Oxford University Press.

Henderson, G. and S. Jin *et al*. 1994. Distribution of Medical Insurance in China. *Social Science of Medicine*. 41(8): 1119-1130.

Henderson, G. and JS. Akin. 1998. Trends in Health Services Utilization in Eight Provinces in China 1989-1993. *Social Science of Medicine*. 47(12): 1957-1971.

Holmer, M. 1984. Tax Policy and the Demand for Health Insurance. *Journal of Health Economics*. 3: 203-221.

Hong, Z. 2002. Rural Cooperative Medical System: for Building of Green Security Line for the Peasants. Kunshan Daily 31.01.2002. http://ksrb.ks.js.cn/20020131/ws/content2.htm. Accessed in June 2003.

Hoogeveen, H. 2003. Evidence on Informal Insurance in Rural Zimbabwe. *Journal of African Economies*. 11(2): 249-278.

Hsiao, W. C. 1995. The Chinese Health Care System: Lessons for Other Nations. *Social Science and Medicine*. 41(8): 1047-1055.

Hsiao, W. C. 2001. Unmet Health Needs of Two Billion: Is Community Financing A Solution? Paper for the Commission on Macroeconomics and Health, World Health Organization.

Jack, W. 1999. *Principles of Health Economics for Developing Countries*. WBI Development Studies. Washington, DC: The World Bank.

Jakab, M. and C. Krishnan. 2001. Community Involvement in Health Care Financing: A Survey of Literature on the Impact, Strengths and Weakness. Health, Nutrition and Population (HNP) Discussion paper. The World Bank.

Jalan, J. and M. Ravallion. 1999. Are the Poor Less Well-Insured? Evidence on Vulnerability to Income Risk in Rural China. *Journal of Development Economics*. 58: 61-81.

Jiang, Y., S. Xiao and Q. Song. 2003. The Impact of Intra-Extended Family Risk Sharing Group on Demand for Health Insurance in Rural Areas: A Pilot Survey Report in Sichuan. *Population and Economy.* 136(1): 74-80.

Juetting, J. 1999. Strengthening Social Security Systems in Rural Areas of Developing Countries. ZEF Discussion Papers on Development 9/1999. Bonn, Germany: Center for Development Research, University of Bonn.

Juetting, J. 2001. The Impact of Health Insurance on the Access to Health Care and Financial Protection in Rural Developing Countries: The Example of Senegal. Health, Nutrition and Population (HNP) Discussion Paper. Washington, DC: The World Bank.

Khan, M., N. Zhu and J.C. Ling. 1997. Designing A Health Insurance Program for Rural Bangladesh: Lessons from the Cooperative Medical System of Taicang County, China. *The Bangladesh Development Studies.* XXV (2): 31-51.

Kimball, M.S. 1990. Precautionary Savings in the Small and in the Large. *Econometrica.* 58(1): 53-73.

Kipnis, A.B. 1997. Producing Guanxi: Sentiment, Self, and Subculture in a North China Village. Durham, N.C.: Duke University Press.

Klose, T. 1999. The Contingent Valuation Method in Health Care. *Health Policy*. 47: 97-123.

Kochar, A. 1999. Smoothing Consumption by Smoothing Income: Hours-of-Work Responses to Idiosyncratic Agricultural Shocks in Rural India. *Review of Economics and Statistics.* 81(2): 50-61.

Kristroem, B. 1990. Valuing Environmental Benefits Using the Contingent Valuation Method: An Econometric Analysis. Umea Economic Studies No. 219, University of Umea.

Lai, L. 2001. Situation and Future Development of Rural Cooperative Medicines System in Qinglong County in Guizhou Province. *China Rural Survey.* 37(1): 48-53. In Chinese.

Ligon, E. 1996. Risk Sharing and Information: Theory and Measurement in Village Economies. Berkeley US: Department of Agriculture and Resource Economics, University of California. October Mimeo.

Liu, J.T, J. Hammitt, J. Wang *et al.* 2000. Mother's Willingness to Pay for Her Own and Her Child 's Health: A Contingent Valuation Study in Taiwan. *Health Economics*. 9: 319-326.

Liu, L. 2001. The Application of Relation among the Immigrant Population—in the Case of Pingjiang Village in Shenzhen. *China's Social Science*. 5: 112-124. In Chinese.

Liu, Y., S. Hu, W. Fu and W.C. Hsiao. 1996. Is Community Financing Necessary and Feasible for Rrural China? *Health Policy*. 38: 155-171.

Liu, Y., W. C. Hsiao and K. Eggleston. 1999. Equity in Health and Health Care: the Chinese Experience. *Social Science & Medicine*. 49(10): 1349-1356.

Loomis, J.B. 1990. Comparative Reliability of Dichotomous Choice and Open-Ended Contingent Valuation Techniques. *Journal of Environmental Economics and Management*. 17: 78-85.

Lund, S. and M. Fafchamps. 1997. Risk Sharing Networks in Rural Philippines. California, US: Department of Economics, Stanford University. Processed.

Ma, Z. 2000. On Rural Primary Health Care System of Chinese Characteristics. *China Health Economics*. 5: 51-52. In Chinese.

Maddala, G.S. 1983. *Limited-Dependent and Qualitative Variables in Econometrics.* Cambridge, UK: Cambridge University Press.

Manning, W. and M.S. Marquis. 1996. Health Insurance: the Trade-off Between Risk Pooling and Moral Hazard. *Journal of Health Economics.* 15: 609-639.

Marquis, M.S. and C.E. Phelps. 1987. Demand for Supplemental Health Insurance. *Economic Inquiry.* 25(2): 299-313.

Martin, D and T. Smeeding. 1985. Eds. *Horizontal Equity, Uncertainty, and Economic Well-being.* Chicago, US: University of Chicago Press.

Maslow, A. 1970. Motivation and Personality. In *Personality Theories.* Boeree, C. G. 1997. http://www.ship.edu/~cgboeree/maslow.html. Accessed in January 2003.

Mathiyazhagan, K. 1998. Willingness to Pay for Rural Health Insurance through Community Participation in India. *International Journal of Health Planning and Management.* 13: 47-67.

Mayers, D. and C.W. Smith. 1983. The Interdependence of Individual Portfolio Decisions and the Demand for Insurance. In *Foundation of Insurance Economics.* Dionne, G., and S. E. Harrington. 1991. Dordrecht, the Netherlands: Kluwer Academic Publisher.

Mead, O. 1992. The Macroeconomic Impact of AIDS in sub-Saharan Africa. The World Bank. http://www.worldbank.org Accessed in May 2001.

Meuwissen, M.P.M. 2000. *Insurance as a Risk Management Tool for European Agriculture.* PhD thesis of Wageningen University, the Netherlands.

McConnell, K.E. 1997. Does altruism undermine existence value? *Journal of Environmental Economics and Management.* 32: 22-37.

MOH. 1995. *China's Health Statistics Yearbook 1995.* Beijing: China's Statistics Press.

MOH. 1999. *China's Health Statistics Yearbook 1999.* Beijing: China's Statistics Press.

MOH. 2002. Statistic Database of Ministry of Health. http://www.moh.gov.cn. Accessed in June 2003.

Moon, M. 1983. Role of the Family in the Economic Well-being of the Elderly. *Gerontologist.* 23 (1): 45-50.

Morduch, J. 1991. Consumption Smoothing Across Space. Cambridge, Massachusetts: Harvard University, Department of Economics. Processed.

Morduch, J. 1999. Between the State and the Market: Can Informal Insurance Patch the Safety Net. *The World Bank Research Observer.* 14(2): 187-207.

Morduch, J. 2002. Consumption Smoothing Across Space: Tests for Village-Level Responses to Risk. WIDER Discussion Paper DP 55/2002. Helsinki Finland: World Institute for Development Economics Research, Insurance Against Poverty Programme.

Morduch, J. and T. Sicular. 1999. Risk and Insurance in Transition: Perspective from the Zouping County, China. In *Market and Community in Economics Development.* Aoki, M. and Y. Hayami. Eds.. New York, US: Oxford University Press.

Mwabu, G. 1990. Financing Health Services in Africa: An Assessment of Alternative Approaches. Working Papers in Population and Human Resources Department June WPS 457. Washington, DC: The World Bank.

Newhouse, J.P. and C.E. Phelps 1976. New Estimates of Price and Income Elasticities of Medical Services. In *the Role of Health Insurance in Health Services Sector.* Rossett, R.N. Ed. US: National Bureau of Economic Research, Inc. 261-320.

NSB. 2001: *China's Statistic Yearbook 2001*. Beijing: China's Statistics Publisher.

------. 2002. *China's Statistic Yearbook 2002*. China's Statistics Publisher

Nugent, J. B. 1990. Old Age Security and the Defense of Social Norms. *Journal of Cross-Cultural Gerontology*. 5: 243-254.

Petretto, A. 1999. Optimal Social Health Insurance with Supplementary Private Insurance. *Journal of Health Economics*. 18(6): 727-745.

Phelps, C. E. 1992. Diffusion of Information in Medical care. *Journal of Economic Perspectives*. Summer 6: 23-42.

Phelps, C.E. 1997. *Health Economics*. 2nd ed. Massachusetts: Addison-Wesley Educational Publishers Inc.

Philips, J.M. 1987. A Comment on Farmer Education and Farm Efficiency: A Survey. *Economic Development and Culture Change*. 35(3): 637-644.

Plateau, J.P. 1997. Mutual Insurance as An Elusive Concept in Traditional Rural Communities. *Journal of Development Studies*. 33(6): 764-796.

Propper, C. 1989. An Econometric Analysis of the Demand for Private Health Insurance in England and Wales. *Applied Economics*. 21: 777-792.

Qiu, Y. and S. Sun. 1999. *Medical Insurance*. Beijing, China: China People University Press

Randall, A. 1997. The NOAA Panel Report: A New Beginning or the End of an Era. *American Journal of Agricultural Economy*. 79: 1489-94.

Ravallion, M. and L. Dearden. 1988. Social Security in A Moral Economy: An Empirical Analysis for Java. *Review of Economics and Statistics*. 70(2): 36-44.

Rejda, G.E. 1998. *Principles of Risk Management and Insurance*. 6th Ed. Massachusetts, US: Addison-Wesley Longman Inc.

Rosenzweig, M.R. and K.I. Wolpin. 1985. Specific Experience, Household Structure, and Intergenerational Transfers: Farm Family Land and Labor Arrangements in Developing Countries. *Quarterly Journal of Economics*. 100 (*Suppl.*): 961-987.

Russell, S, J. Fox-Rushby, D.C. Arhin. 1995. Willingness and Ability to Pay for Health Care: A Selection of Methods and Issues. *Health policy and planning*. 10(1): 94-101

Sahn, D.E. 2000. *The Demand for Health Care Services in Rural Tanzania*. Paper for International Conference—Health Systems Financing in Low-Income African and Asian Countries.

Sauerborn, R., A. Sdams, and M. Hien. 1996a. Household Strategies to Cope with the Economic Costs of Illness. *Social Science and Medicine*. 43(3): 291-301.

Sauerborn, R., A. Nougtara and M. Hien *et al*. 1996b. Seasonal Variations of the Household Costs of Illness in Burkina Faso. *Social Science and Medicine*. 43: 281-290.

Sawada, Y. 1997. Income Risks, Gender, and Human Capital Investment in Rural Pakistan. Stanford: Department of Economics, Stanford University. Mimeo.

Schoeni, R.F. 1993. Private Inter-household Transfers of Money and Time: New Empirical Evidence. *Rand document*. Document No: DRU-443-NICHD.

Schultz, T. W. 1961. Investment in Human Capital. *American Review*. 51(1).

Shaw, R.P. and C.C. Griffin. 1995. *Financing Health Care in Sub-Saharan Africa through User Fees and Insurance*. The World Bank.

Sine, J. 1994. *Demand for Episodes of Care in the China Health Insurance Experiment*. Dissertation of Rand Graduate School, Rand, USA.

Sköld, M. 1998. *Poverty and Health: Who Lives, Who dies, Who cares?* Geneva: World Health Organization.

SSB. 2002. *Sichuan Statistics Yearbook 2002*. Beijing: China's Statistics Publisher.

Tan, X. 1997. Implement Farmer Hospitalization Health Insurance and Establish Farmer Health Security System. *China Rural Health Administration.* 17(9): 5-6.

Tang, S.1997. The Changing Role of the Township Health Centers in the Context of Economic Reform in China. *IDS. Bulletin* 28(1): 39-47

Thompson, R. and R. Carr-Hill. 2000 *Barriers to Health Care in Kazakhstan: Findings from A Household Survey*. Paper for International Conference—Health Systems Financing in Low-Income African and Asian Countries.

Townsend, R. 1994. Risk and Insurance in Village India. *Econometrica* 62: 539-92.

Train, K. 2003. *Discrete Choice Methods with Simulation.* Cambridge, UK: Cambridge University Press.

United Nation Development Program (UNDP). 2002. Human Development Report 2002. http://hdr.undp.org/reports/view_reports.cfm?year=2002. Accessed in July 2003.

Von Braun, J. and P. Webb. 1989. The Impact of New Crop Technology on the Agricultural Division of Labor in West Africa Setting. *EDCC*. 37(3): 513-534.

Von Braun, J., T. Teklu and P. Webb. 1999. Famine in Africa: Causes, Responses, and Prevention. Baltimore, Maryland: *The Johns Hopkins University Press.*

Walraven, G. 1996. Willingness to Pay for District Hospital Services in Rural Tanzania. *Health Policy Plan.* 11: 428-437.

Wang, W. 2002. The Impact of Information and Telecommunication Technology on Farmers in China. Frankfurt am Main: Peter Lang, Europaeischer Verlag der Wissenschaft.

Waters, H.R. 2000. Measuring Equity in Access to Health Care. *Social Science & Medicine.* 51: 599-612.

Weinberger, K. and J. Juetting. 2000. The Role of Local Organization in Risk Management: Some Evidence from Rural Chad. *Quarterly Journal of International Agriculture*. 39: 281-298.

WHO. 1995. *The World Health Report 1995: Bridging the gaps.* Geneva: World Health Organization.

WHO. 1996. *Equity in Health and Health Care: A WHO/SIDA Initiative*. Geneva: World Health Organization.

WHO. 1998. Progress in Reproductive Health Research. WHO News 45, part 1: www.who.int/reproductive-health/hrp/45/news45_1.en.html. Accessed in April 2003.

WHO. 2000. The *World Health Report 2000—Health System: Improving Performance*. Geneva: World Health Organization.

Wiesmann, D. and J. Juetting. 2000. *The Demand for Health Insurance in Rural Areas of Low-Income Countries—A Theoretical Framework.* Paper for international conference—health systems financing in low-income African and Asian countries.

World Bank. 1980. *World Development Report 1980*. Washington, DC: The World Bank.

World Bank. 1984. *The Health Sector in China—A World Bank Country Study*. Washington, DC: The World Bank.

World Bank. 1993. *World Development Report-Investing in Health*. The World Bank.

World Bank. 1997. *Financing Health Care: Issues and Options for China.* Washington, DC:

The World Bank.

World Bank. 2002. *China at a Glance*. Washington, DC: The World Bank.

World Bank. 2003. *World Development Report 2003: Sustainable Development in a Dynamic World.* Washington DC: The World Bank

Wu, X. and Research Team on Chinese Commercial Health Insurance. 2001. *Study on Chinese Commercial Health Insurance*. China's Finance Publisher. In Chinese.

Wu, Y. 1997. *China's Health Sector in Transition: Resources, Demand and Reforms. Health Policy*. 39: 137-152.

Xinhua Agency. 2002. Establishment of China's Rural Cooperative Medical System. November 12. www.ah.xinhua.org. Accessed in May 2003.

Yin, X., Q. Wang and L. Jin. 1998. Investigation and Reflection on Rural Cooperative Medical System in Nanjing. *Jiangsu Preventive Medicine.* 3:59-60.

Yip, W.C., H. Wang, and Y. Liu. 1998. The Impact of Health Financing on Demand for Health Care in Rural China. *Health Policy and Planning*. 13(3): 311-322.

Young, D.L. 1984: *Risk Concepts and Measures for Decision Analysis*. In *Risk Management in Agriculture*. Barry, P.J. (ed.): 31-42. Ames, Iowa: Iowa State University Press.

Yu, H. 2002. Analysis of Barrier to the Formation of China's Pharmaceuticals' Administrative System. http://www.unirule.org.cn/academia/inst0205.htm. Accessed in May 2002.

Zeller, M. 1999. Towards Enhancing the Role of Micro-Finance for the Safety Nets of the Poor. ZEF Discussion Papers on Development 19/1999. Bonn, Germany: Center for Development Research, University of Bonn.

Zeller, M., S. Gertrud, J. von Braun and F. Heidhues. 1997. *Rural Finance for Food Security for the Poor: Implications for Research and Policy*. Food Policy Review 4. International Food Policy Research Institute, Washington, D. C.

Zethraeus, N. 1998. Willingness to Pay for Hormone Replacement Therapy. *Health Economics*. 7: 31-38.

Zhang, Z., J. Zhang and J. Zhang. 1997. Investigation on Strengthening the Rural Health Organization and Developing Rural Health Care Scheme in Poverty-Stricken Districts. *China's Rural Health Administration*. 17(1): 59-60. In Chinese.

Zheng, B, C. Zhou, Y. Xia and *et al.* 1998. The Rural Cooperative Medical System Must Introduce the Insurance Mechanism: A Survey on RCMS in 12 Counties, Jiangsu Province. *Jiangsu Health Administration.* 42(2): 108-111. In Chinese.

Zhong, G., J. Zhuo and Z. Zheng *et al.* 2002. A Survey on Children's Planning Immunization in 28 Counties of Guangxi. *Guangxi Preventive Medicine*. 8(1): 17-18. In Chinese.

Zhou, C. 1999. Survey and Economic Analysis of Rural Cooperative Medical Scheme. *China's Rural Health Administration*. 19(1): 38-39. In Chinese.

Zhou, S. *et al.* 1991. The Need for Re-establishing Cooperative Medical System. *Chinese Rural Health Administration. 65(2): 78-81.*

Zhu, L. 2000. Government and the Institutional Options of Rural Basic Health Care Financing. http://newhome.baoding.cn.net/~jzf/tg12.htm. Accessed in 2001.

Zhu, N., A. Ling, J. Shen, J.Lane and S. Hu. Factors Associated with the Decline of the Cooperative Medical System and Bare Foot Doctors in Rural China. *Bulletin of the World Health Organization*. 67: 431-441.

Appendix A Major health insurance policies or schemes in rural China

Table A-1 Main private health insurance (PHI) products of China Life*

Type of insurance		Name of insurance policy	Way of sale	Period of insurance	Way of indemnity
Individual insurance	Hospitalization type	Hospitalization allowance insurance	Major insurance	1 year	Allowance
		Additional hospitalization allowance insurance (I)	Additional insurance	1 year	Allowance + quota
		Additional hospitalization allowance insurance (II)	Additional	1 year	Expense
		Additional inpatient health insurance	Additional	1y	Expense
	Dread disease type	Dread diseases benefit lifetime*	Major	Life time	Quota
		Dread diseases benefit survival-life*	Major	Periodical	Quota
		Green shade for life—major diseases protector	Major	Periodical	Quota
		Emergency medical insurance •	Major		
		Female disease benefit type A and type B	Major		
		Additional three specific dread diseases benefit	Additional		
	Accident type	Additional accident medical insurance	Additional	1 year	Expense + co-payment
		Additional accident medical expenses cover*	Additional	1 year	Expense + quota
		Additional accident living allowances cover	Additional	1 year	Allowance
Group insurance	Hospitalization type	Pupil and kindergarten kid hospitalization supplementary medical insurance*	Major	1 year	Expense
		Pupil and kindergarten kid additional inpatient care coverage	Additional	1 year	Expense
	Major diseases	Additional group comprehensive health insurance	Additional	1 year	Expense
		Group health insurance	Major	1 year	Expense
		Group female health insurance	Major	1 yeat	Expense
		Green shade for life—major disease protector	Major	Periodical	Quota
		Emergency medical insurance	Major		
	Accident type	Accident injury care benefit	Major		
		Pupil and kindergarten additional accident medical insurance*	Additional	1 year	Expense

Note: * marks the often-encountered health insurance policies in rural areas.

Table A-2 Often encountered forms of RCMS and their features in China

		Welfare type of RCMS	Risk type of RCMS	Risk-welfare type of RCMS	Comprehensive community health insurance
Objectives		Ensure primary and preventive care	Pool the risk of major diseases	Sum of welfare and risk types	Pool risk, ensure primary & preventive care, health maintenance & hygiene
Technical design	Premium level	Low, 1% of average net income per capita last year	Medium, 1-1.5% of average net income per capita last year	High, 2% of average net income per capita last year	High, 2-3% of average net income per capita last year
	Source of benefits fund	Individuals and collective (village)	Individuals and collective(village or town)	Individuals and collective(village, or town)	Individuals and collective (town or county)
	Benefits package	Outpatient services & some preventive care	Inpatient services & some preventive care	Outpatient services, vision care & dental care	Out- & inpatient services, prevention, health maintenance and environmental hygiene
	Way of benefit	Either expenses of medicines or of treatment	Expenses of medicines and of curing of inpatient care	Both expenses of medicines and of treatment	Both expenses of medicines and of treatment
Management	Sponsor and administrator	Initiated by village and administered by town or both by town	Initiated and administered by town	Initiated and administered by town	Initiated by town and administered by county or both by county
	Distribution of fund	Medical fund 80%, preventive fund 15%, reserve and management fund 5%	Medical fund 85%, preventive fund 10%, reserve fund and management fund 5%	Medical fund 75%, preventive fund 10%, risk fund 10%, and reserve and management fund 5%	Medical fund 75%, preventive fund 10%, risk fund 10%, and reserve and management fund 5%
Attributes	Advantages	Affordability, applicability to the poverty region. Easiness to raise premium	Affordability, applicability to middle income areas.	Strong ability to resist risk of major disease, affordability and applicability to rich areas	High coverage, strong ability to resist risk of major disease, affordability and applicability to rich areas
	Disadvantages	Low coverage and no capacity of resisting major risk	Difficulty in raising money and in management	Less beneficiaries	Less beneficiaries

Source: compiled from documents of Jian Ge health bureau 1998, Mimeograph.

Table A-3 Often encountered prepayment schemes in China

	Immunization prepayment scheme	Perinatal prepayment scheme	Child health care prepayment scheme
Target population	All the children at 0-7 years old without contraindication of vaccine	Pregnant and lying-in women	All children at 0-7 year old
Money prepaid	40-60 Yuan/person.	40-48 Yuan/person	40 Yuan/person
Services covered	6 statutory vaccines preventing 8 diseases	12 antenatal visits, 3 postnatal visits, postpartum examination of mother and child 42 days after delivery.	4 check-ups at age of 1, 2 check-ups from 2 to 3 years old, 1 check-ups from 4 to 7 years old. Some health instructions.
Managerial organization of funds	Anti-epidemic station and township health center	Maternity and child health care station and township health center	Maternity and child health station and township health center
Distribution of funds	65% consulting fee, 15% equipment fee, 10% bonus and 10% compensation reserve	70% consulting fee, 10-15% compensation reserve, 5-10% equipment fee and 5% bonus	The same as perinatal scheme
Compensation for failure	20-200 Yuan for immunization failure	50-2000 Yuan for health care failure	50-2000 Yuan for health care failure

Source: compiled on the basis of National Congress 1994; Sichuan Health Bureau 1994

Appendix B Robustness test of the impact of RCMS on health status at county level

Table B-1 Estimates of health production function with continuous measure of RCMS: Log-linear model

	IMR		MMR	
	Coefficient	Standard Error	Coefficient	Standard Error
(Constant)	4.046***	0.866	4.234	2.677
Log average annual net income per farmer	-0.703***	0.222	-0.503	0.297
Log licensed rural doctor per capita	-0.102	0.073	0.173	0.225
Log ratio of population with clean drinking water	0.055	0.146	0.019	0.452
Log public spending on health care per capita	-0.005	0.202	0.035	0.623
Log welfare expenditure per capita	-0.224**	0.103	-0.165	0.320
Log ratio of RCMS member	0.214	0.363	0.098	1.123

Source: computed from ZEF survey 2002

Appendix C Calculation of the impact of RCMS on IMR over time—An after-before comparative analysis of the whole country

Because the data available are ratios of the urban to the rural, we can rewrite it

Eq.0-1 $\lg R_{Hb} = \lg \frac{H_{bu}}{H_{br}} = \lg H_{bu} - \lg H_{br}$

Eq.0-2 $\lg H_{bu} = c + \alpha \lg I_{bu} + \beta \lg S_{bu}$

Eq.0-3 $\lg H_{br} = c + \alpha \lg I_{br} + \beta \lg S_{br}$

where R_{Hb} is referred to as health indicator ratio of the urban to the rural before the breakdown of RCMS, the subscript *bu* refers to 'before' and 'urban', subscript *br* refers to 'before' and 'rural'. R_{Ib} and R_{Sb} refer to income ratio and resources mobilized ratio, respectively; the subscript *ua* and *ra* refer to 'urban and after' and 'rural and after". Then $\lg R_{Hb}$ equals equation (0-2) minus (0-3)

Eq.0-4) $\lg R_{Hb} = \alpha(\lg I_{bu} - \lg I_{br}) + \beta(\lg S_{bu} - \lg S_{br})$

$$= \alpha \lg R_{Ib} + \beta \lg R_{Sb}$$

∴ Eq.0-5) $\lg R_{Sb} = \lg \frac{S_{ub}}{S_{rb}} = \frac{\lg R_{Hb} - \lg R_{Ib}}{\beta}$

likewise

Eq.0-6) $\lg R_{Sa} = \lg \frac{S_{ua}}{S_{ra}} = \frac{\lg R_{Ha} - \lg R_{Ia}}{\beta}$

we then compare the R_{Sa} and R_{Sb},

$$\lg \frac{R_{Sa}}{R_{Sb}} = \lg R_{Sa} - \lg R_{Sb} = \frac{1}{\beta}[(\lg R_{Ha} - \alpha \lg R_{Ia}) - (\lg R_{Hb} - \alpha \lg R_{Ib})]$$

Eq.0-7) $= \frac{1}{\beta}(\lg \frac{R_{Ha}}{R_{Hb}} - \alpha \lg \frac{R_{Ia}}{R_{Ib}})$

then we judge the sign of representation equation by Eq.(0-7) replacing four ratios with value in Table 5-4.

$$\frac{R_{Ha}}{R_{Hb}} = \frac{3.35}{1.65} = 2.03$$

$\frac{R_{Ia}}{R_{Ib}} = \frac{2.57}{2.40} = 1.07$ Moreover, $0 < \alpha < 1$

∴ Equation (5-9) > 0

$$\lg \frac{R_{Sa}}{R_{Sb}} > 0 \Rightarrow \frac{R_{Sa}}{R_{Sb}} > 1$$

$$R_{Sa} > R_{Sb}\text{, namely, } \frac{S_{au}}{S_{ar}} > \frac{S_{bu}}{S_{br}}$$

We know that health care financing mechanism in urban areas has not changed greatly, so it can be assumed that S_{au} equals S_{bu} approximately. But health care financing mechanism in rural areas has changed dramatically. Thus the conclusion may be drawn now that S_{ar} is less than S_{br}. That is to say, the resources mobilized for the rural before the breakdown of RCMS were larger than that after the breakdown. Rural cooperative medical scheme did play a big role in narrowing the gap of health indicators between the urban and the rural population.

Appendix D WTP for a complete benefits package with *no* co-payment

Figure D-1 Variation of WTP with income at different mother's education

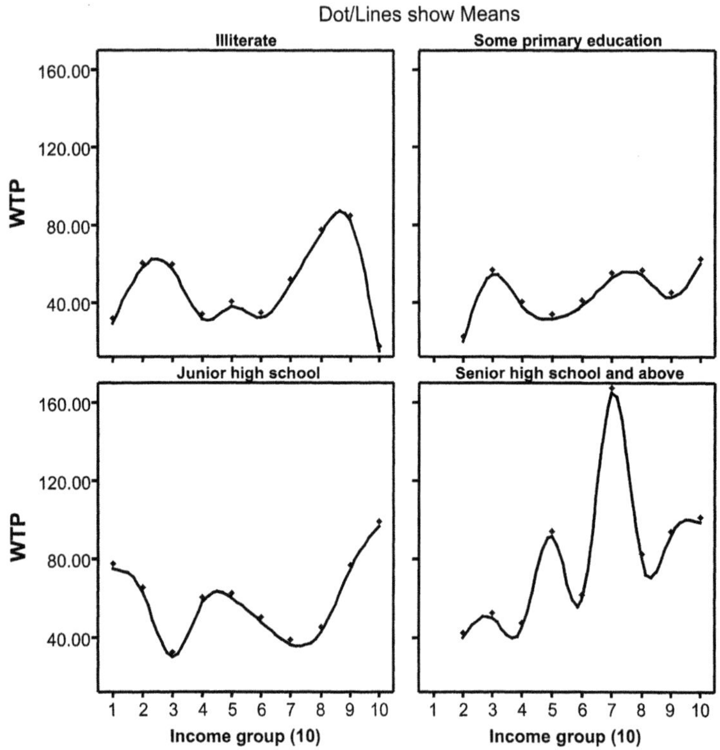

Figure D-2 Variation of WTP with income at different level of land area

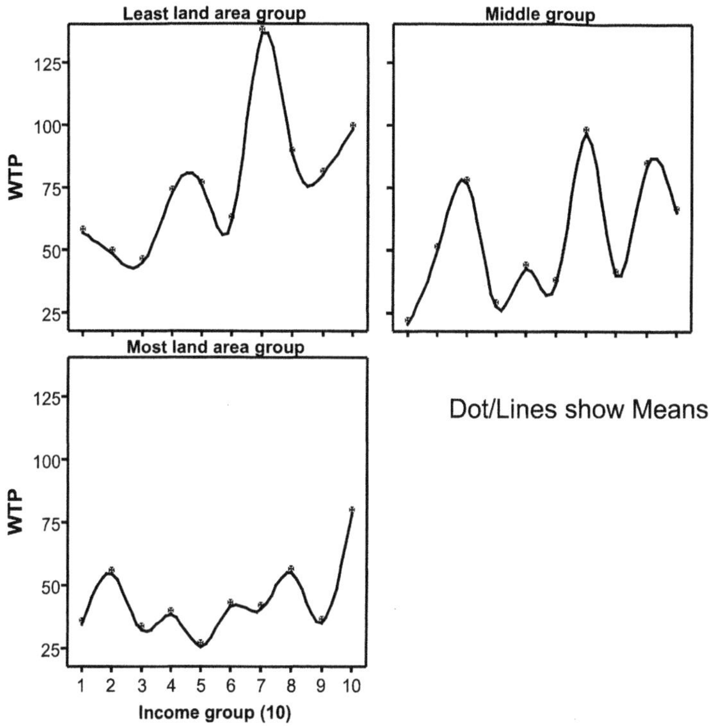

Development Economics and Policy

Series edited by Franz Heidhues and Joachim von Braun

Band 1 Andrea Fadani: Agricultural Price Policy and Export and Food Production in Cameroon. A Farming Systems Analysis of Pricing Policies. The Case of Coffee-Based Farming Systems. 1999.

Band 2 Heike Michelsen: Auswirkungen der Währungsunion auf den Strukturanpassungsprozeß der Länder der afrikanischen Franc-Zone. 1995.

Band 3 Stephan Bea: Direktinvestitionen in Entwicklungsländern. Auswirkungen von Stabilisierungsmaßnahmen und Strukturreformen in Mexiko. 1995.

Band 4 Franz Heidhues / François Kamajou: Agricultural Policy Analysis – Proceedings of an International Seminar, held at the University of Dschang, Cameroon on May 26 and 27 1994, funded by the European Union under the Science and Technology Program (STD). 1996.

Band 5 Elke M. Förster: Protection or Liberalization? A Policy Analysis of the Korean Beef Sector. 1996.

Band 6 Gertrud Schrieder: The Role of Rural Finance for Food Security of the Poor in Cameroon. 1996.

Band 7 Nestor R. Ahoyo Adjovi: Economie des Systèmes de Production intégrant la Culture de Riz au Sud du Bénin: Potentialités, Contraintes et Perspectives. 1996.

Band 8 Jenny Müller: Income Distribution in the Agricultural Sector of Thailand. Empirical Analysis and Policy Options. 1996.

Band 9 Michael Brüntrup: Agricultural Price Policy and its Impact on Production, Income, Employment and the Adoption of Innovations. A Farming Systems Based Analysis of Cotton Policy in Northern Benin. 1997.

Band 10 Justin Bomda: Déterminants de l'Epargne et du Crédit, et leurs Implications pour le Développement du Système Financier Rural au Cameroun. 1998.

Band 11 John M. Msuya: Nutrition Improvement Projects in Tanzania: Implementation, Determinants of Performance, and Policy Implications. 1998.

Band 12 Andreas Neef: Auswirkungen von Bodenrechtswandel auf Ressourcennutzung und wirtschaftliches Verhalten von Kleinbauern in Niger und Benin. 1999.

Band 13 Susanna Wolf (ed.): The Future of EU-ACP Relations. 1999.

Band 14 Franz Heidhues / Gertrud Schrieder (eds.): Romania – Rural Finance in Transition Economies. 2000.

Band 15 Katinka Weinberger: Women's Participation. An Economic Analysis in Rural Chad and Pakistan. 2000.

Band 16 Christof Batzlen: Migration and Economic Development. Remittances and Investments in South Asia: A Case Study of Pakistan. 2000.

Band 17 Matin Qaim: Potential Impacts of Crop Biotechnology in Developing Countries. 2000.

Band 18 Jean Senahoun: Programmes d'ajustement structurel, sécurité alimentaire et durabilité agricole. Une approche d'analyse intégrée, appliquée au Bénin. 2001.

Band 19 Torsten Feldbrügge: Economics of Emergency Relief Management in Developing Countries. With Case Studies on Food Relief in Angola and Mozambique. 2001.

Band 20 Claudia Ringler: Optimal Allocation and Use of Water Resources in the Mekong River Basin: Multi-Country and Intersectoral Analyses. 2001.

Band 21 Arnim Kuhn: Handelskosten und regionale (Des-)Integration. Russlands Agrarmärkte in der Transformation. 2001.

Band 22 Ortrun Anne Gronski: Stock Markets and Economic Growth. Evidence from South Africa. 2001.

Band 23 Patrick Webb / Katinka Weinberger (eds.): Women Farmers. Enhancing Rights, Recognition and Productivity. 2001.

Band 24 Mingzhi Sheng: Lebensmittelkonsum und -konsumtrends in China. Eine empirische Analyse auf der Basis ökonometrischer Nachfragemodelle. 2002.

Band 25 Maria Iskandarani: Economics of Household Water Security in Jordan. 2002.

Band 26 Romeo Bertolini: Telecommunication Services in Sub-Saharan Africa. An Analysis of Access and Use in the Southern Volta Region in Ghana. 2002.

Band 27 Dietrich Müller-Falcke: Use and Impact of Information and Communication Technologies in Developing Countries' Small Businesses. Evidence from Indian Small Scale Industry. 2002.

Band 28 Wolfram Erhardt: Financial Markets for Small Enterprises in Urban and Rural Northern Thailand. Empirical Analysis on the Demand for and Supply of Financial Services, with Particular Emphasis on the Determinants of Credit Access and Borrower Transaction Costs. 2002.

Band 29 Wensheng Wang: The Impact of Information and Communication Technologies on Farm Households in China. 2002.

Band 30 Shyamal K. Chowdhury: Institutional and Welfare Aspects of the Provision and Use of Information and Communication Technologies in the Rural Areas of Bangladesh and Peru. 2002.

Band 31 Annette Luibrand: Transition in Vietnam. Impact of the Rural Reform Process on an Ethnic Minority. 2002.

Band 32 Felix Ankomah Asante: Economic Analysis of Decentralisation in Rural Ghana. 2003.

Band 33 Chodechai Suwanaporn: Determinants of Bank Lending in Thailand: An Empirical Examination for the Years 1992 to 1996. 2003.

Band 34 Abay Asfaw: Costs of Illness, Demand for Medical Care, and the Prospect of Community Health Insurance Schemes in the Rural Areas of Ethiopia. 2003.

Band 35 Gi-Soon Song: The Impact of Information and Communication Technologies (ICTs) on Rural Households. A Holistic Approach Applied to the Case of Lao People's Democratic Republic. 2003.

Band 36 Daniela Lohlein: An Economic Analysis of Public Good Provision in Rural Russia. The Case of Education and Health Care. 2003.

Band 37 Johannes Woelcke. Bio-Economics of Sustainable Land Management in Uganda. 2003.

Band 38 Susanne M. Ziemek: The Economics of Volunteer Labor Supply. An Application to Countries of a Different Development Level. 2003.

Band 39 Doris Wiesmann: An International Nutrition Index. Concept and Analyses of Food Insecurity and Undernutrition at Country Levels. 2004.

Band 40 Isaac Osei-Akoto: The Economics of Rural Health Insurance. The Effects of Formal and Informal Risk-Sharing Shemes in Ghana. 2004.

Band 41 Yuansheng Jiang: Health Insurance Demand and Health Risk Management in Rural China. 2004.

www.peterlang.de

Abay Asfaw

Costs of Illness, Demand for Medical Care, and the Prospect of Community Health Insurance Schemes in the Rural Areas of Ethiopia

Frankfurt am Main, Berlin, Bern, Bruxelles, New York, Oxford, Wien, 2003.
XVII, 239 pp., num. fig. and tab.
Development Economics and Policy.
Edited by Franz Heidhues and Joachim von Braun. Vol. 34
ISBN 3-631-50553-1 / US-ISBN 0-8204-6095-8 · pb € 45.50*

After the 'Health for All' declaration of the Alma Ata conference, many countries in the developing world have been trying to provide basic health services to their population. However, it has become increasingly clear that due to market failure, state failure, and other reasons, the conventional sources of finance could not solve the health problem of developing countries, particularly that of the rural population and the high-risk groups. As a result, there is a worldwide initiative to find additional sources of finance which are broad based, sustainable, and suitable to the socio-economic situations of the rural population. This study assesses the prospect and potential role of community health insurance schemes in the provision of basic health care services and in generating sustainable finance for the health sector of Ethiopia. It starts by investigating the costs of illness, health care demand behaviour, and the capacity of existing risk-sharing arrangements in protecting households against health shocks. Then, it thoroughly analyses the willingness of households to pay and the prospect of potential community health insurance schemes in the rural areas of Ethiopia.

Contents: New approach in measuring costs of illness in the rural areas of developing countries · Evaluating the capacity of existing risk-sharing institutions in protecting households against health shocks · Analysing the health care demand behaviour of households

Frankfurt am Main · Berlin · Bern · Bruxelles · New York · Oxford · Wien
Distribution: Verlag Peter Lang AG
Moosstr. 1, CH-2542 Pieterlen
Telefax 00 41 (0) 32 / 376 17 27

*The €-price includes German tax rate
Prices are subject to change without notice

Homepage http://www.peterlang.de

Health Insurance Demand and Health Risk Management in Rural China

DEVELOPMENT ECONOMICS AND POLICY

Series edited by Franz Heidhues and Joachim von Braun

Vol. 41

PETER LANG

Frankfurt am Main · Berlin · Bern · Bruxelles · New York · Oxford · Wien

Health Insurance Demand and Health Risk Management in Rural China

Yuansheng Jiang

PETER LANG
Europäischer Verlag der Wissenschaften

Bibliographic Information published by Die Deutsche Bibliothek
Die Deutsche Bibliothek lists this publication in the Deutsche Nationalbibliografie; detailed bibliographic data is available in the internet at <http://dnb.ddb.de>.

Zugl.: Bonn, Univ., Diss., 2004

D 98
ISSN 0948-1338
ISBN 3-631-52624-5
US-ISBN 0-8204-7312-X

www.peterlang.de